ESSENTIAL
BUILDING SERVICES
AND EQUIPMENT

ESSENTIAL BUILDING SERVICES AND EQUIPMENT

F. Hall
HNC, DipCIOB, CEd, CGFTC, RP
First Prize and Silver Medal,
City and Guilds of London Institute

BUTTERWORTH
HEINEMANN

Butterworth-Heinemann
Linacre House, Jordan Hill, Oxford OX2 8DP
225 Wildwood Avenue, Woburn, MA 01801-2041
A division of Reed Educational and Professional Publishing

R A member of the Reed Elsevier plc group

OXFORD   AUCKLAND   BOSTON
JOHANNESBURG   MELBOURNE   NEW DELHI

First published 1988
Reprtined 1989, 1990, 1991, 1993, 1994
Second edition 1995
Reprinted 1996, 1998, 1999

British Library Cataloguing in Publication Data
Hall, F.
Essential building services and equipment.
1. Buildings - Environmental engineering
I Title
696   TH6021

ISBN 0 7506 2556 2

Printed and bound in Great Britain by MPG Books Ltd, Bodmin, Cornwall

CONTENTS

PREFACE

The cost of building services and equipment can amount to between 40 and 60 per cent of the total cost of a modern complex building. Architects, surveyors and builders – everyone concerned with the building construction has to have a sound knowledge of the subject.

This book has been designed to cover, concisely and clearly, a wide range of services, so as to provide a handy reference work for all building industry personnel, from the craftsman to the higher technician and the site manager. It will also be useful to students preparing for the examinations of the Royal Institution of Chartered Surveyors, the Chartered Institute of Building, the City and Guilds of London Institute, the Business and Technology Education Council (BTEC) and the National Vocational Qualifications (NVQs).

The services covered in the book include hot and cold water systems, drainage, heating, heat pumps, heat recovery, air conditioning, ventilation, refuse and sewage disposal, electric, gas and oil installations, fire control, lighting and lifts.

The emphasis of the book is on diagrams that illustrate the principles and show the methods used and the standards that must be met. The written description is kept to the minimum necessary for a clear explanation, in the belief that it is in studying the diagrams that the reader will get a firm grip of method and detail.

In this second edition, additional work on the important topic of the conservation of energy has been added in the appendix and, where necessary, the text has been revised. Gas ignition devices and small-bore pumped waste systems have also been added in the appendix.

I should like to give a very special thanks to my late wife for her patience and understanding during the preparation of the first edition.

F. Hall

1 COLD WATER AND SUPPLY SYSTEMS

RAIN CYCLE — SOURCES OF WATER SUPPLY

Surface sources Lakes, streams, rivers, reservoirs, run off from roofs and paved areas.

Underground sources Shallow wells, deep wells, artesian wells, artesian springs, land springs.

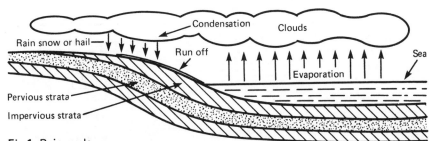

Fig 1 Rain cycle

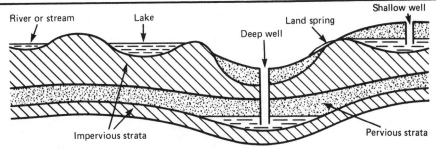

Fig 2 Surface and normal underground supplies

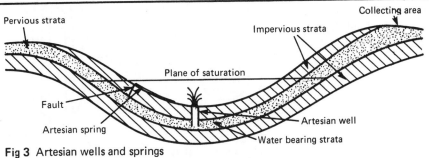

Fig 3 Artesian wells and springs

1

FILTRATION OF WATER

Notes
1   Rate of filtration 4 to 12 m³ per m² per hour.
2   To back wash valve A is closed and valves B and C opened.
3   Compressed air clears the sand of dirt.
4   Diameter = 2·4 m.

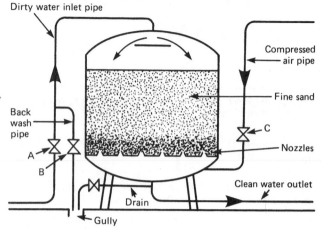

Fig 4 Pressure filter

1   Rate of filtration 0·2 to 1·15 m³ per m² per hour.
2   The filter takes up a lot more space than the pressure filter.
3   Top layer of sand requires cleaning.

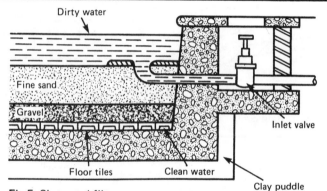

Fig 5 Slow sand filter

1   The unglazed porcelain cylinder will arrest very fine particles of dirt and even micro-organisms.
2   The cylinder can be removed and sterilised in boiling water for 10 min.

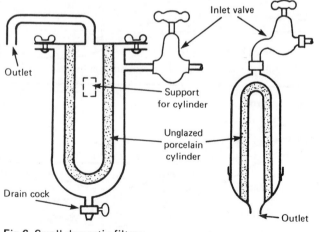

Fig 6 Small domestic filters

STERILISATION AND SOFTENING

Notes

1   Water used for drinking must be sterilised.

2   Chlorine is generally used for this purpose which destroys organic matter.

3   The chlorine is normally added after filtration and the residual chlorine must be kept between 0·1 and 0·3 ppm.

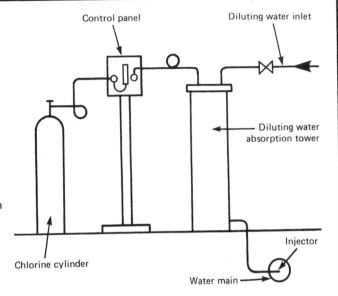

Fig 7 Sterilisation of water by use of chlorine injected into the water

1   Sodium zeolites exchange their sodium base for calcium or magnesium bases in the water.

2   Sodium zeolite plus calcium carbonate or sulphate becomes calcium zeolite plus sodium carbonate or sulphate.

3   To regenerate salt is added; calcium zeolite plus sodium chloride becomes sodium zeolite plus calcium chloride which is flushed away to a gully.

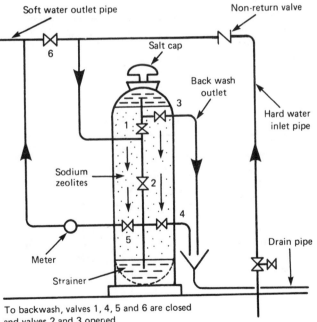

To backwash, valves 1, 4, 5 and 6 are closed and valves 2 and 3 opened

Fig 8 Softening of hard water by the base-exchange process

STORAGE AND DISTRIBUTION OF WATER

Notes

1   The water from upland gathering grounds is impounded in a dam.

2   From this point the water is filtered and chlorinated before it serves a town or village at lower level.

3   The method saves pumping.

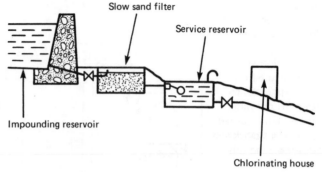

Fig 9 Gravitational distribution

1   The water from the river is pumped into a settlement tank.

2   From this point the water is filtered and chlorinated.

3   Pumping is more expensive than gravitational distribution.

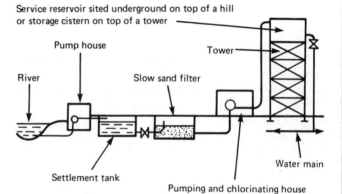

Fig 10 Pumped distribution

1   The water mains supplying the town or village may be in the form of a grid.

2   This grid provides a ring circuit and each section can be isolated.

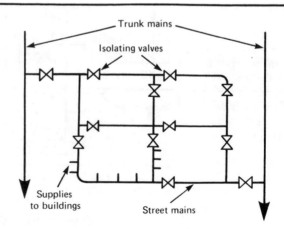

Fig 11 Ring main distribution

VALVES USED FOR WATER — 1

Notes

1 The globe type stop valve is used to control the flow of water at high pressure through pipework.

2 In order to shut off the water the crutch is turned slowly in a clockwise direction and this reduces the risk of water hammer.

3 The gate or sluice valve is used to control the flow of water at low pressure.

4 The valve is closed by turning the wheel in a clockwise direction and the valve offers far less resistance to flow than the globe valve.

5 The drain valve is used for draining pipework, vessels, and boilers.

6 For temperatures of up to 100°C the valves are usually made from brass and for higher temperatures gun metal is used. Brass contains 50% zinc and 50% copper. Gun metal contains 85% copper, 5% zinc, 5% lead and 5% tin.

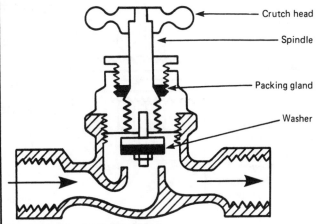

Crutch head
Spindle
Packing gland
Washer

Fig 12 Stop valve (globe type)

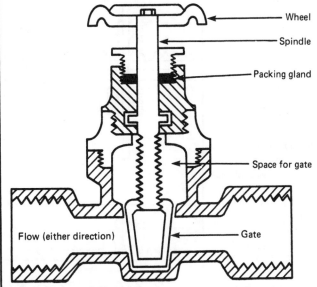

Wheel
Spindle
Packing gland
Space for gate
Flow (either direction)
Gate

Fig 13 Gate or sluice valve

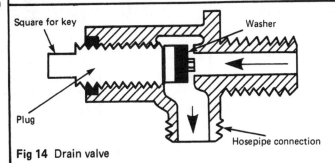

Square for key
Washer
Plug
Hosepipe connection

Fig 14 Drain valve

5

VALVES USED FOR WATER — 2

Notes

1   Ball valves are used to allow water to flow into a cistern and to shut off the supply when the correct water level has been reached. Various types are used.

2   The diaphragm ball valve reduces noise and there is less sticking of the moving parts.

3   Most Water Supply Companies will not permit a silencer pipe unless it is well above the highest water level. This is to prevent siphonage of water from the cistern into the water main (back siphonage).

4   The piston on the Portsmouth and Croydon valves moves horizontally and vertically respectively.

5   High, medium and low pressure ball valves must be capable of closing against a pressure 1379, 689·5 and 276 kPa respectively.

6   The diameters of the nozzles are reduced as the pressure increases.

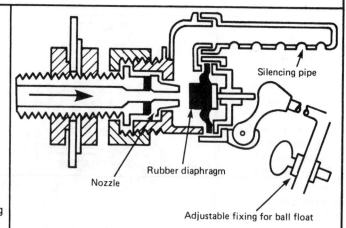

Fig 15 Diaphragm ball valve

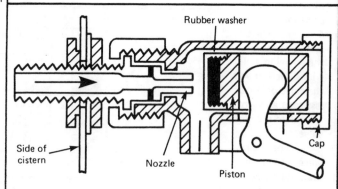

Fig 16 Portsmouth ball valve BS 1212

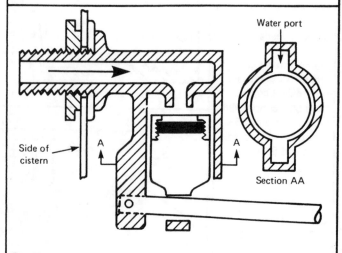

Fig 17 Croydon ball valve

TAPS USED FOR WATER

Notes

1   The pillar tap is used to supply water to basins, baths, bidets and sinks.

2   Combined hot and cold pillar taps are available with a combined fixed or swivel outlet.

3   The outlet of these taps must have separate waterways for the hot and cold supply, known as 'bi-flow' outlet.

4   The bib tap is for fixing on the wall about 150 mm above a sanitary appliance.

5   The 'Supatap' bib tap permits the change of washer without shutting off the water supply. They are very easy to turn on and off and are suitable for the disabled.

6   When mixer taps are used the outlet must have a bi-flow. This will prevent the possibility of mixing of the hot and cold water before discharging through the open end.

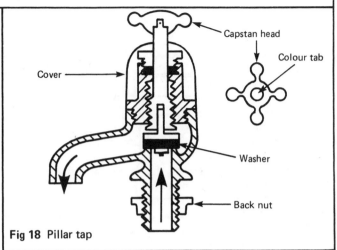

Fig 18 Pillar tap

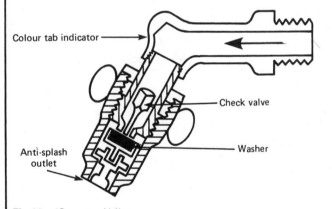

Fig 19 'Supatap' bib tap

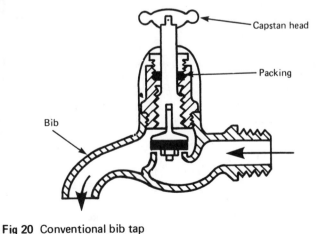

Fig 20 Conventional bib tap

7

JOINTS ON WATER PIPES

Notes

1 Copper pipes may be jointed by special fittings or by bronze welding.

2 Non-manipulative compression joints are used above ground.

3 Manipulative compression joints are used below ground and prevent the pipes pulling out of the joint.

4 The push-fit joint is made from polybutylene and saves on labour.

5 The capillary joint contains soft solder rings and the solder flows to form the joint when the fitting is heated.

6 The Talbot joint is used on polythene pipes.

7 Threaded joints on steel pipes may be jointed by non-toxic jointing paste or by use of polytetrafluoroethylene (PTFE) tape.

8 A taper thread on the pipe will help to ensure a water-tight joint.

9 The union joint permits slight deflection without leakage.

10 Lead pipes cause lead poisoning and must not be used.

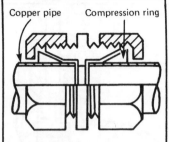

Fig 21 Non-manipulative compression joint on copper pipes

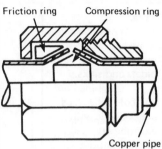

Fig 22 Manipulative compression joint on copper pipes

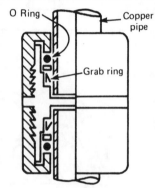

Fig 23 Acorn push-fit joint on copper pipes

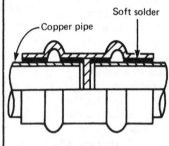

When the fitting is heated solder flows

Fig 24 Soft soldered capillary joint on copper pipes

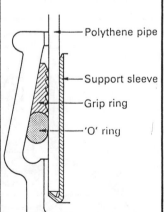

Fig 25 The Talbot push-fit joint on polythene pipes

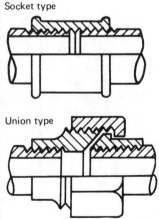

Fig 26 Screwed joints on mild steel pipes

CONNECTION TO WATER MAIN

Notes

1   At least seven days notice in writing is required for connection to the water main.

2   The water main is drilled and tapped by a special apparatus with the water left on.

3   A plug valve is left on the main and a communication pipe connected to it.

4   A goose neck is formed on the pipe to relieve stresses on the pipe and valve.

5   The pipe is laid on a bed of sand 50 mm thick.

6   A stop valve is fitted on the pipe inside a box with a hinged cover.

7   The underground service pipe should be 'snaked' to allow for settlement of the pipe.

8   If plastic pipe is used during hot weather the pipe, when laid in a cool trench, will contract quite considerably and, the 'snaked' pipe will allow for this contraction and prevent stress on the pipe.

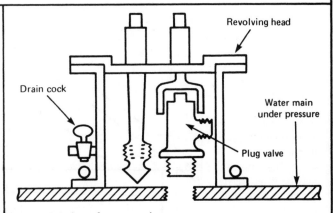

Fig 27 Tapping of water main

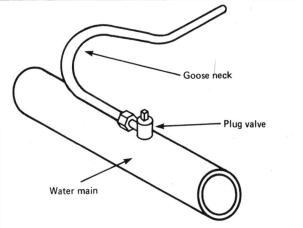

Fig 28 View of water main connection

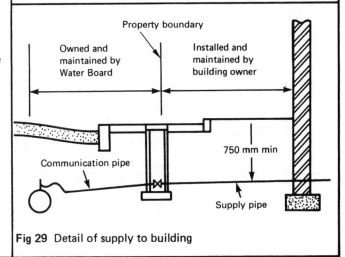

Fig 29 Detail of supply to building

DIRECT SYSTEM OF COLD WATER SUPPLY

Notes

1 Good pressure on the water main is required.

2 Pipework is reduced to a minimum.

3 A feed cistern 114 litre min. capacity is required.

4 The cistern may be fitted inside the airing cupboard.

5 Drinking water is available at every draw-off point.

6 All pipes fixed to fall to drain valves.

7 Pipes to be fixed 750 mm min. from outside walls.

8 Valves may be fitted to isolate sections of pipework.

9 Greater risk of water contamination by back siphonage.

10 Back siphonage occurs when there is a large demand on the water main and water is drawn back from pipes inside the building. Foul water from a submerged inlet will be siphoned back into the main.

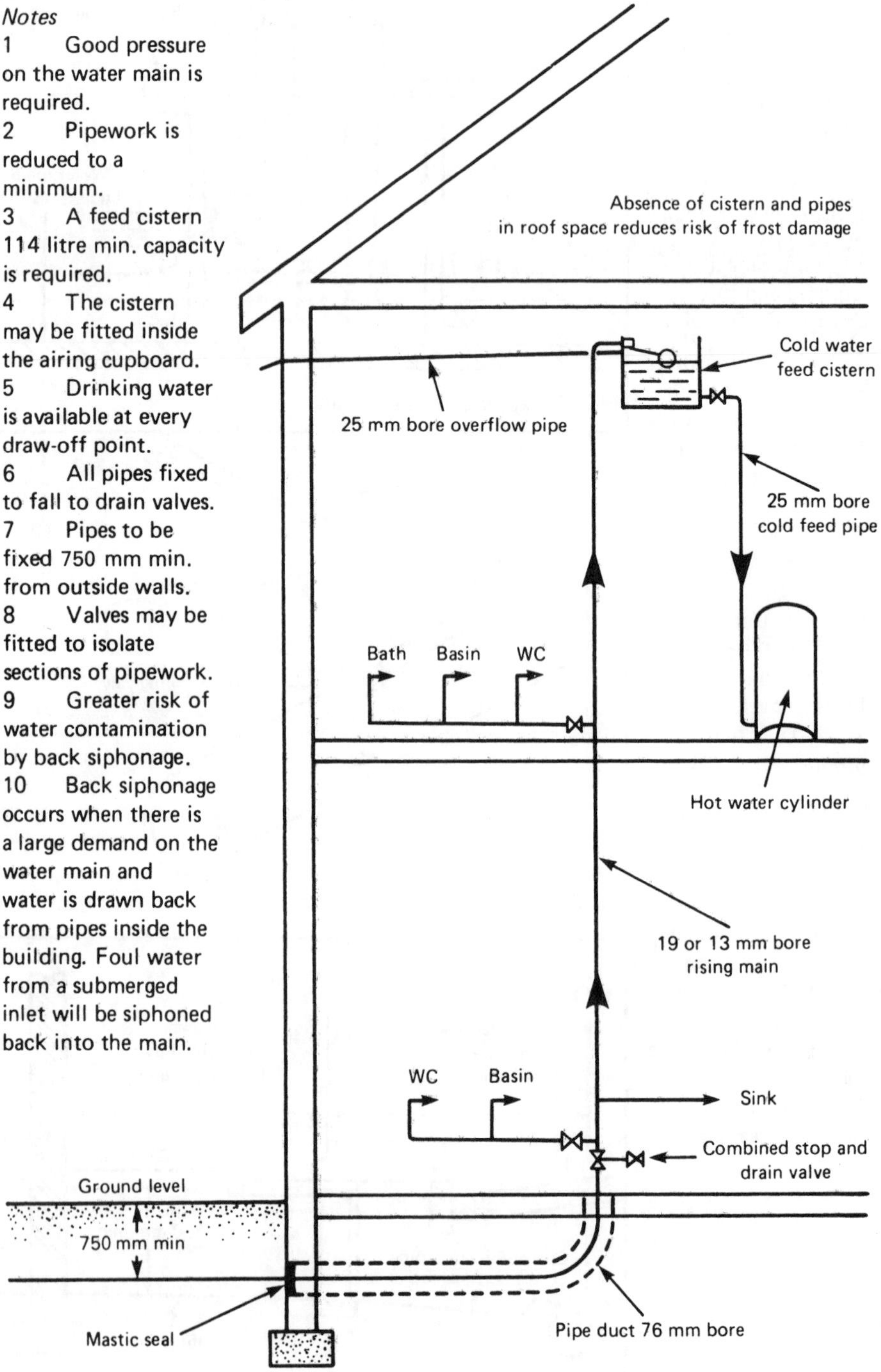

Absence of cistern and pipes in roof space reduces risk of frost damage

Cold water feed cistern

25 mm bore overflow pipe

25 mm bore cold feed pipe

Bath   Basin   WC

Hot water cylinder

19 or 13 mm bore rising main

WC   Basin

Sink

Combined stop and drain valve

Ground level

750 mm min

Mastic seal

Pipe duct 76 mm bore

Fig 30

INDIRECT SYSTEM OF COLD WATER SUPPLY

Notes

1   In the indirect system of cold water supply, drinking water is normally allowed only at the sink or drinking fountain.

2   A cold water storage cistern 227 litre min. capacity is required and this larger cistern requires fixing in the roof space.

3   The system requires more pipework than the direct system and is more expensive to install.

4   Fewer fittings connected direct to the main reduces risk of pollution by back-siphonage.

5   Fittings supplied from cistern may have a reduced water pressure, which reduces noise and wear on valves.

6   A shower may be easily supplied from the cistern.

7   The system provides a large reserve of water.

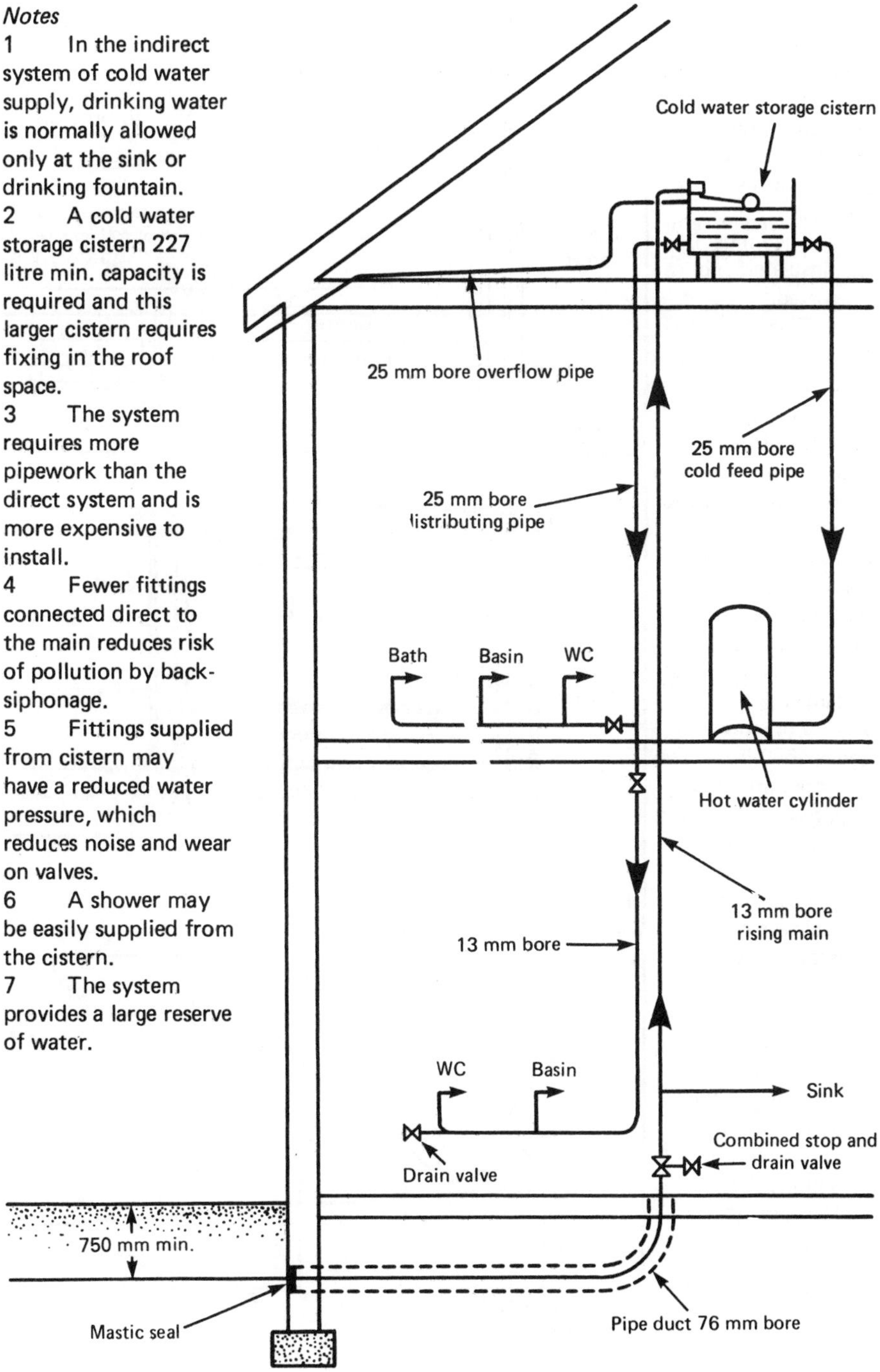

Cold water storage cistern

25 mm bore overflow pipe

25 mm bore cold feed pipe

25 mm bore distributing pipe

Bath   Basin   WC

Hot water cylinder

13 mm bore rising main

13 mm bore

WC   Basin   Sink

Combined stop and drain valve

Drain valve

750 mm min.

Mastic seal

Pipe duct 76 mm bore

Fig 31

COLD WATER STORAGE CISTERNS

Notes

1 Cisterns can be made from galvanised mild steel, polypropylene, asbestos cement or glass reinforced plastics.

2 They must be well insulated and supported.

3 They must be provided with a dust proof cover.

4 For large buildings cisterns are housed inside a cistern room on the roof top.

5 The cistern room must be well insulated and the air temperature must not fall below 0°C.

6 The base must be tanked.

7 For storage above 4500 litres, two or more cisterns are required.

8 The cisterns must be inter-connected so that each cistern may be isolated for cleaning, repair or renewal.

9 The addition of a drain valve on each cistern will permit the draining of the cisterns without the need to open the taps inside the building, or siphon the water out.

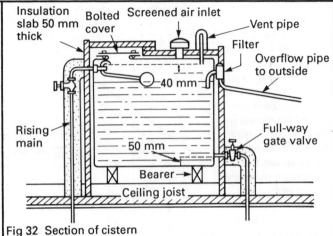

Fig 32 Section of cistern

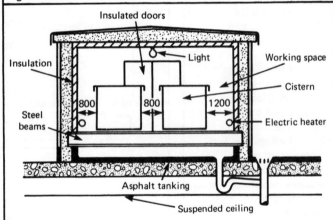

Fig 33 Detail of cistern room

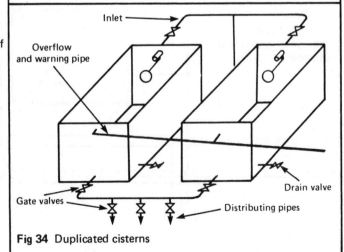

Fig 34 Duplicated cisterns

BOOSTED COLD WATER SYSTEM — 1

Notes

1   If the pressure on the main will not force water up to the upper floor boosting is required.

2   A break cistern prevents pumping direct from the main.

3   A break-pressure cistern lowers the pressure on the lower floor and the head of water above the valves should not exceed 30 m.

4   The drinking water header pipe supplies drinking water to the upper floors and when this empties the pipeline switch cuts in the duty pump.

5   The lower float switch protects the pumps. If there is an interruption of the mains supply the float eventually falls and switches off the duty pump.

6   Isolation valves are fitted on the various sections of pipework to facilitate maintenance work.

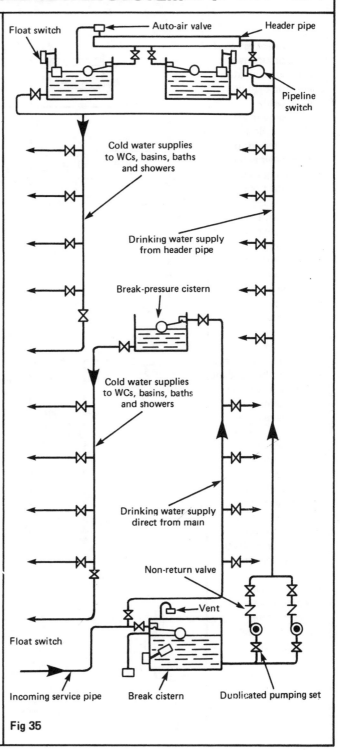

Float switch

Auto-air valve

Header pipe

Pipeline switch

Cold water supplies to WCs, basins, baths and showers

Drinking water supply from header pipe

Break-pressure cistern

Cold water supplies to WCs, basins, baths and showers

Drinking water supply direct from main

Non-return valve

Vent

Float switch

Incoming service pipe

Break cistern

Duplicated pumping set

Fig 35

13

BOOSTED COLD WATER SYSTEM — 2

Notes

1 As an alternative to the drinking water header pipe an auto-pneumatic cylinder may be used.

2 Compressed air in the cylinder forces water up to the ball valves and the drinking water taps on the upper floors.

3 When the cylinder is emptied a low-pressure switch cuts in the duty pump.

4 When the pump refills the cylinder a high-pressure switch cuts out the duty pump.

5 Some air is absorbed by the water and when this happens a float switch cuts in the air compressor. A filter is required on the compressed air pipeline to prevent dirt entering the cylinder.

6 The break pressure cisterns may be supplied either from the storage cisterns at roof level, or from the rising main.

7 A pressure reducing valve is sometimes used instead of a break pressure cistern.

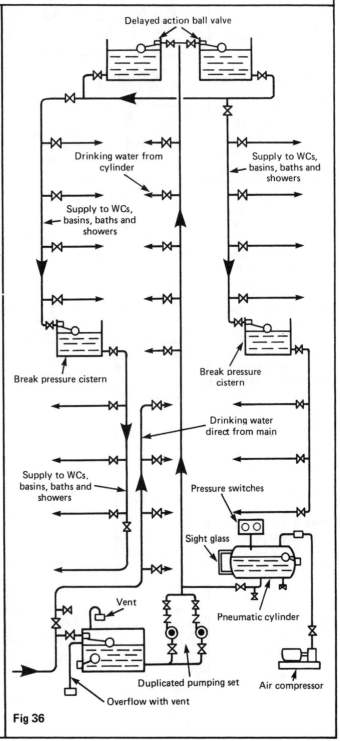

Delayed action ball valve

Drinking water from cylinder

Supply to WCs, basins, baths and showers

Supply to WCs, basins, baths and showers

Break pressure cistern

Break pressure cistern

Drinking water direct from main

Supply to WCs, basins, baths and showers

Pressure switches

Sight glass

Vent

Pneumatic cylinder

Duplicated pumping set

Air compressor

Overflow with vent

Fig 36

2 HOT WATER SUPPLY SYSTEMS

UNVENTED HOT WATER STORAGE SYSTEM

Notes
1 The Building Regulations 1991 permit the installation of unvented hot water storage systems.
2 The system saves the cost of a storage and expansion cistern and associated pipework.
3 Before the system is installed a check should be made to ensure that there is adequate pressure on the main.
4 The sealed primary circuit can be pumped or can circulate by natural convection.
5 It is easy to install a shower with a good water pressure.
6 The system should be in the form of a proprietary unit or package approved by a member body of the European Organization for Technical Approvals (EOTA).

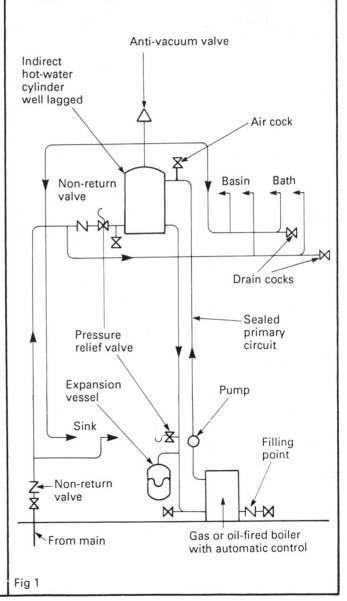

Fig 1

INDIRECT SYSTEM OF HOT WATER SUPPLY

Notes

1   The system is used in hard water areas to prevent scaling of boiler and pipes.

2   This is because the water in the boiler and primary circuit is not drawn off through the taps which prevents fresh water entering.

3   The system is also used when heating is combined with the system and prevents water from the radiators passing out of the hot water taps.

4   The system costs more than the direct system but requires less maintenance.

5   The boiling temperature may be up to 70°C.

6   It is not advisable to fit a stop valve on the primary cold feed pipe. If a valve is fitted it may be accidentally shut off, thus leaving the primary circuit without a supply of cold water.

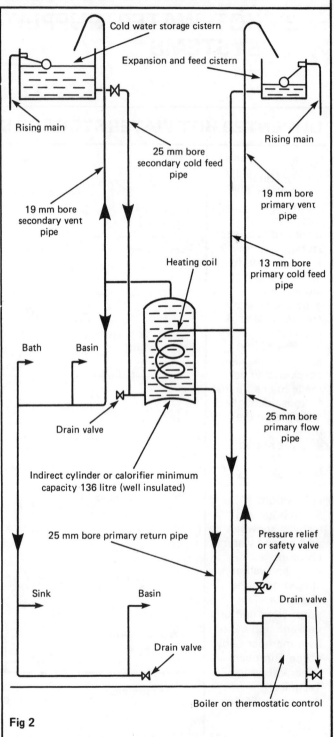

Cold water storage cistern

Expansion and feed cistern

Rising main

Rising main

25 mm bore secondary cold feed pipe

19 mm bore primary vent pipe

19 mm bore secondary vent pipe

13 mm bore primary cold feed pipe

Heating coil

Bath   Basin

Drain valve

25 mm bore primary flow pipe

Indirect cylinder or calorifier minimum capacity 136 litre (well insulated)

25 mm bore primary return pipe

Pressure relief or safety valve

Sink   Basin

Drain valve

Drain valve

Drain valve

Boiler on thermostatic control

Fig 2

PRIMATIC CYLINDER — DUPLICATED PLANT

Notes

1. An indirect hot water system may be installed using a Primatic cylinder.
2. The heat exchanger in the cylinder has three air locks which prevents mixing of the primary and secondary waters.
3. An expansion and feed cistern primary cold feed and primary vent pipe are not required.
4. The system saves a good deal of cost.
5. Duplicated plant is required to ensure a supply of hot water at all times.
6. Each boiler calorifier and pump may be isolated for repair or renewal.
7. If it is required to drain one of the boilers for repair or replacement the isolating valves are closed and the three-way vent valve turned so that the boiler is open to atmosphere. The boiler may be then emptied through the drain valve.

Sf = Secondary flow pipe
Pf = Primary flow pipe
Pr = Primary return pipe
He = Heat exchanger
Cf = Cold feed pipe

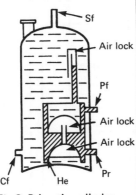

Fig 3 Primatic cylinder

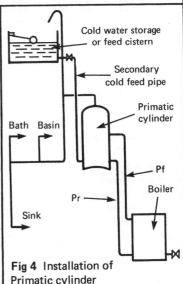

Fig 4 Installation of Primatic cylinder

Pcf = Primary cold feed pipe
Vv = Vent valve
Scf = Secondary cold feed pipe
Pv = Primary vent pipe
Sv = Secondary vent pipe
Nrv = Non-return valve
Sf = Secondary flow pipe
Sr = Secondary return pipe
Dps = Duplicated pumps
3 Wvv = 3-way vent valve
Dv = Drain valve

Fig 5 Duplicated plant

INDIRECT HOT WATER SYSTEM FOR A THREE-STOREY BUILDING

Notes

1   For larger buildings a secondary circuit will be required which may be pumped.

2   One of the valves near the pump should be motorised and shut off with the pump and boiler when hot water is not required.

3   The calorifier must be placed close to the boiler to save on heat losses and to provide a quick heat recovery.

4   The boiler, calorifier and hot water pipes must be insulated.

5   A heating circuit may be included.

6   If the heating system is large it may be necessary to install separate boilers for heating and hot water supply.

7   The secondary circuit will operate by thermo-siphonage and the pump omitted. The pump however creates a better circulation of water.

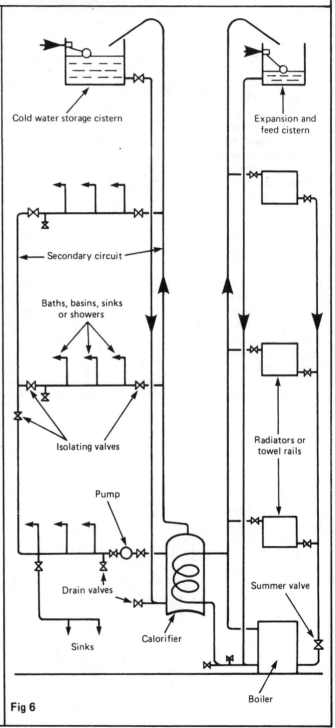

Cold water storage cistern

Expansion and feed cistern

Secondary circuit

Baths, basins, sinks or showers

Isolating valves

Radiators or towel rails

Pump

Drain valves

Summer valve

Sinks

Calorifier

Boiler

Fig 6

'SEALED' INDIRECT HOT WATER SYSTEM FOR HIGH RISE

Notes

1   The maximum head of water above the draw-off points is 30 m and a break-pressure cistern is required.

2   If bidets are installed the hot water supply to them must be by a separate pipe from the calorifiers.

3   The boiler and calorifiers must be able to withstand the water pressure.

4   Head tanks ensure a good delivery of water on the upper floors.

5   The 'sealed' primary circuit saves on pipework and the expansion and feed cistern.

6   The expansion vessel takes up the expansion of water in the primary circuit.

7   In order to reduce heat losses the pipes, calorifiers, head tanks and boiler must be well insluated.

8   The cold water storage and break pressure cisterns must be insulated to prevent damage by frost or heat gain.

9   The Building Regulations permit the use of sealed or unvented systems.

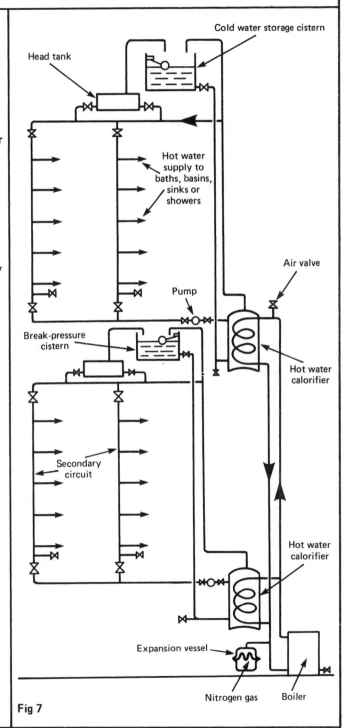

Fig 7

ELECTRIC WATER HEATERS — 1

Notes

1   Electric immersion heaters may be inserted inside the conventional cylinder.

2   Self-contained open outlet heaters may be fitted above basins, baths or sinks.

3   Self-contained cistern type heaters can be used to supply hot water to several sanitary appliances.

4   The heating of towel rails by electrically heated water must be avoided.

5   The thermostat should be set at a maximum of 60°C.

6   Hot water pipes must be as short as possible especially the supply to the sink.

7   The immersion heater must be electrically earthed and the cable supplying the heater must be adequate for the power load.

8   The heaters may be supplied from a low-tariff off-peak electricity supply.

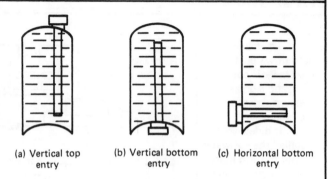

(a) Vertical top entry

(b) Vertical bottom entry

(c) Horizontal bottom entry

Fig 8 Positions of electric immersion heater inside cylinder

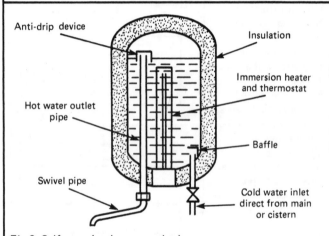

Anti-drip device

Insulation

Immersion heater and thermostat

Hot water outlet pipe

Baffle

Swivel pipe

Cold water inlet direct from main or cistern

Fig 9 Self-contained open outlet heater

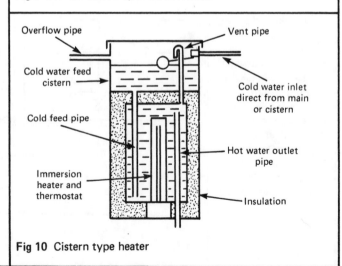

Overflow pipe

Vent pipe

Cold water feed cistern

Cold water inlet direct from main or cistern

Cold feed pipe

Hot water outlet pipe

Immersion heater and thermostat

Insulation

Fig 10 Cistern type heater

ELECTRIC WATER HEATERS — 2

Notes

1   The cistern-type heater must be fitted well above the hot water draw-off taps.

2   The pressure type cistern may have a small heater at the top for basin and sink use and a larger heater at the bottom for bath use.

3   The pressure heater must be supplied with cold water from a cistern and may be fitted close to the sink.

4   The instantaneous heater is used to supply hot water to a basin, sink or shower.

5   A thermostat regulates the water temperature and also the flow of cold water into the heater.

6   An instantaneous shower heater has an electric loading of 6 kW and will give a continuous supply of warm water of about 3 litres per minute.

7   An instantaneous handwash heater has an electric loading of 3 kW and will give a continuous supply of warm water of about 1·4 litres per minute.

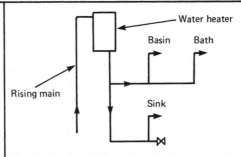

Fig 11 Installation of electric cistern-type heater

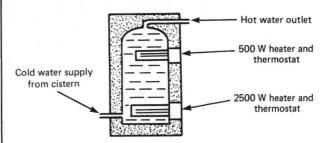

Fig 12 Pressure-type electric water heater

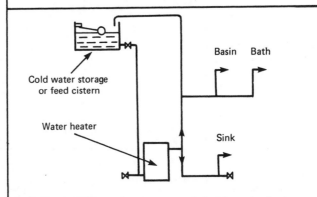

Fig 13 Installation of pressure-type electric water heater

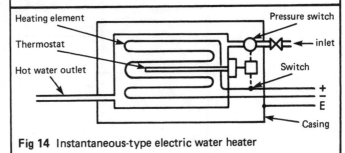

Fig 14 Instantaneous-type electric water heater

GAS WATER HEATERS — 1

Notes

1	The cold water flows through a venturi which produces a differential pressure across the diaphragm and the gas valve is opened.

2	A pilot flame lights the burner and the water is heated as it passes through the heat exchanger.

3	The heater may be supplied direct from the main or a cold water cistern.

4	A multi-point heater may supply several draw-off points.

5	A gas circulator is used to heat water in a storage cylinder.

6	The 3-way valve is turned to circulate water through the higher or lower return pipe and thus save in the use of fuel.

7	The circulator may be installed in the kitchen with vertical flow and return pipes connected to the cylinder fitted inside the airing cupboard.

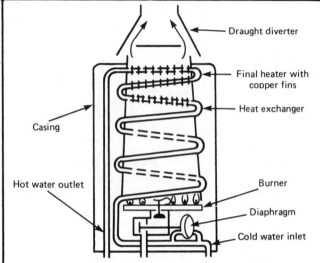

Fig 15 Instantaneous gas water heater

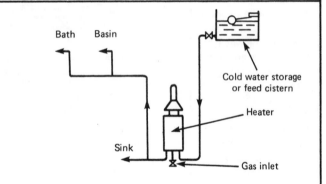

Fig 16 Installation of instantaneous gas water heater

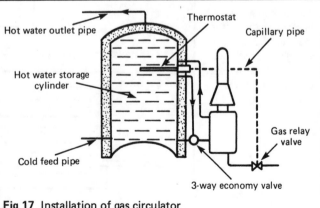

Fig 17 Installation of gas circulator

GAS WATER HEATERS — 2

Notes

1 The storage type gas water heater is a self-contained unit which is quicker to install than the circulator.

2 A small open outlet type storage heater may be used to supply hot water to a sink or basin.

3 Larger type storage heaters will supply hot water to several fittings and these are called multipoint heaters.

4 The heater may be installed in a house to supply bath, basin, sink and shower.

5 It may also be installed in up to 3-storey flats supplied with cold water from one cistern; a vent pipe on the cold feed will prevent siphonage.

6 To prevent hot water from the heaters on the upper floors flowing down to the heater on the ground floor the branch connection on the cold feed pipe must be above the heaters.

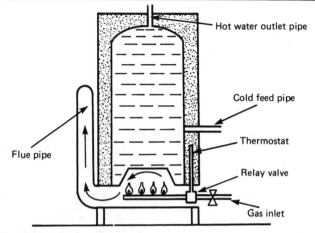

Fig 18 Detail of gas storage heater

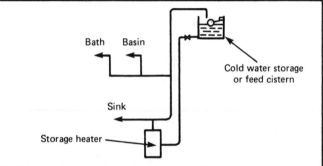

Fig 19 Installation of gas storage heater for a house

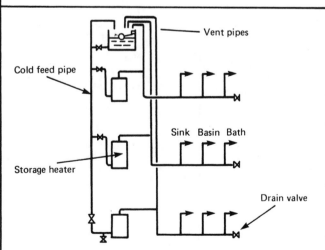

Fig 20 Installation of gas storage heaters for three storey flats (electric pressure heaters may be similarly installed)

SOLAR HEATING OF WATER

Notes

1   Solar energy will save about 40% of the energy required to heat the water.

2   The collector should be 4–6 m² in area; fixed at an angle of 40° to the horizontal and face towards south.

3   The solar cylinder should hold 200 litres of water which will be raised to 60°C.

4   The cylinders and pipes must be insulated.

5   A non-toxic anti-freeze must be added to the water in the solar system.

6   The pump is switched on when the temperature of the water at point X exceeds that at point Y by 2–3°C.

7   If required, the solar cylinder and the conventional cylinder may be fitted on the same level.

8   A combined solar and conventional cylinder is obtainable.

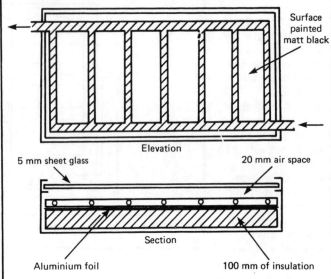

Surface painted matt black

Elevation

5 mm sheet glass

20 mm air space

Section

Aluminium foil

100 mm of insulation

Fig 21 Detail of flat plate solar collector

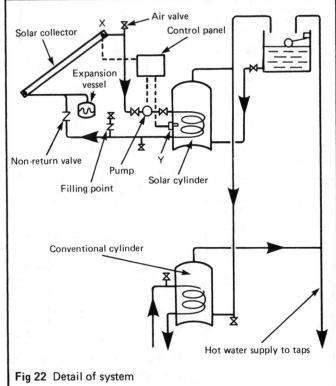

Solar collector

X

Air valve

Control panel

Expansion vessel

Non-return valve

Pump

Filling point

Y

Solar cylinder

Conventional cylinder

Hot water supply to taps

Fig 22 Detail of system

3   HEATING SYSTEMS – 1

LOW TEMP. HOT WATER HEATING SYSTEMS

Notes

1   In a low temperature hot water heating system the maximum temperature of the water is about 80°C.

2   The system may be 'open' with a cistern or 'sealed' with an expansion vessel.

3   The system may operate by gravity or a pump.

4   Pumped circuits use smaller bore pipes.

5   The type of system will depend upon the type of building.

6   A ring circuit is used for a single-storey building.

7   The drop and ladder systems are used for two or more storeys.

8   The drop system is self venting and the radiators will not become air locked.

9   The systems may operate by thermo-siphonage and the pump omitted. The pump however, provides better circulation and smaller pipe may be used.

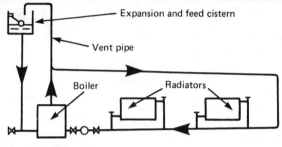

Fig 1   One-pipe ring

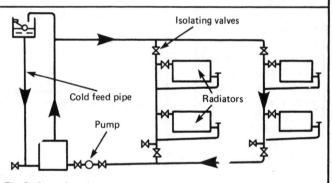

Fig 2   One-pipe drop

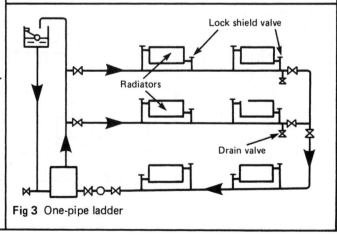

Fig 3   One-pipe ladder

25

LOW TEMP. HOT WATER HEATING SYSTEMS

Notes

1   The one-pipe parallel is useful when the return pipe can be fixed in the suspended ceiling.

2   The disadvantage with all one-pipe systems is the greater cooling of the radiators due to the cooler water from radiators at the start of the circuit being passed to the radiators at the end of the circuit.

3   This cooling of the radiators is reduced by the two-pipe system which requires less regulating.

4   The two-pipe reverse-return or equal travel system requires the least regulating and this is because the lengths of 'travel' to each radiator are equal.

5   The systems may operate by thermo-siphonage and the pump omitted. The pump however, provides better circulation and the pipe sizes may be smaller.

6   The pump may be fitted on either the flow or return pipe.

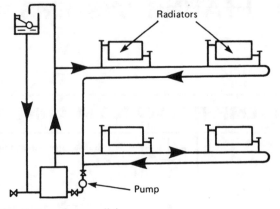

Fig 4 One-pipe parallel

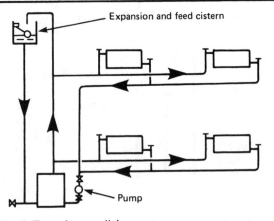

Fig 5 Two-pipe parallel

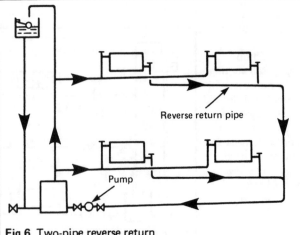

Fig 6 Two-pipe reverse return

LOW TEMP. HOT WATER HEATING SYSTEMS

Notes

1 The two-pipe upfeed system is used when it is not practicable to have horizontal pipes at high level, the main flow and return pipes may be fixed along the corridor inside a floor duct.

2 The two-pipe drop when it is practicable to have the flow-pipe at high level, the radiators are self venting.

3 The two-pipe high-level return system is particularly useful when the installation is in an existing building having a solid ground floor, the system may save cutting away the concrete for the pipes.

4 The systems may operate by thermo-siphonage and the pump omitted. The pump however provides better circulation and smaller pipes may be used.

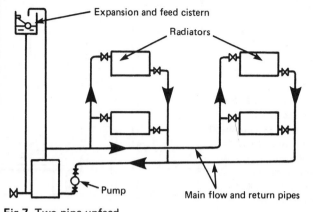

Fig 7 Two-pipe upfeed

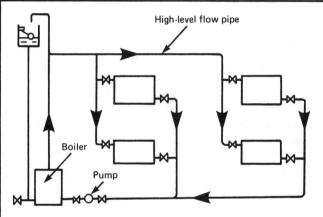

Fig 8 Two-pipe drop

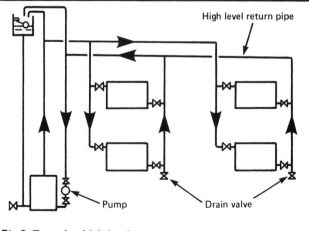

Fig 9 Two-pipe high-level return

EXPANSION VESSELS IN HEATING SYSTEMS

Notes

1 In any water heating system provision must be made for expansion of the water.

2 An expansion and feed cistern provides an expansion space for normal boiler firing plus an additional $33\frac{1}{3}\%$ for high boiler firing is provided.

3 In sealed systems an expansion vessel is required and this should be connected to the return pipe.

4 The use of an expansion vessel saves on labour and pipe work. There is also less risk of air entering the system and less corrosion.

5 The nitrogen gas is pressurised so as to produce a minimum water pressure at the highest point on the heating system of 10 kPa (approx 1 m head of water). The maximum pressure of gas is 300 kPa.

6 Some manufacturers use air for the cushion. Nitrogen however, is an inert gas and will not cause corrosion of the vessel.

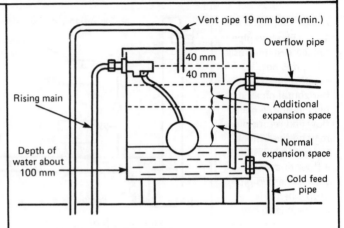

Fig 10 Expansion and feed cistern

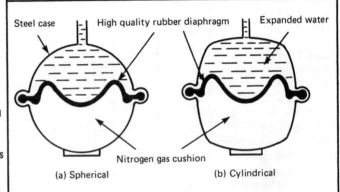

Fig 11 Diaphragm expansion vessels

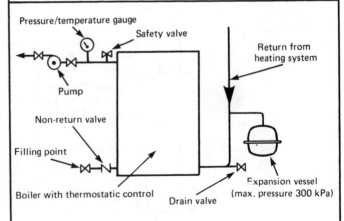

Fig 12 Installation of expansion vessel

LOW TEMP. HOT WATER HEATING SYSTEMS

Notes

1 The small bore heating system uses 15 or 22 mm o.d. (outside diameter) copper tube.

2 The circuit to the cylinder may be pumped or circulated by natural convection.

3 A one- or a two-pipe heating system may be used.

4 The micro-bore system uses 6, 8, 10 or 12 mm o.d. soft copper tubes.

5 A 22 or 28 mm o.d. manifold supplies the radiators through these 'micro-bore' copper pipes.

6 The system may be open or sealed as shown in Fig 14.

7 The micro-bore pipes are easy to install and conceal.

8 The systems will *not* operate by thermo-siphonage and a pump is therefore always required. This is because very small pipes are used which offer high resistance to the flow of water.

9 The room thermostat should be fitted 1·2 m and 1·5 m above the floor.

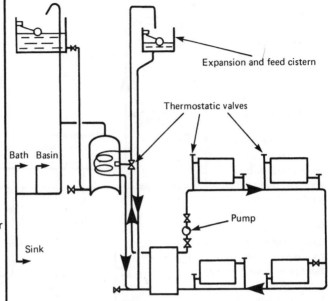

Fig 13 Small-bore system

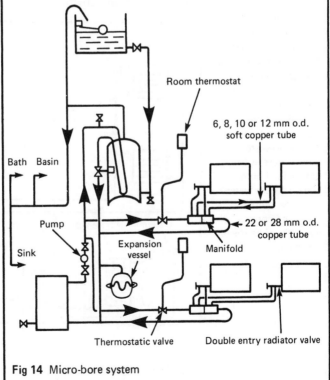

Fig 14 Micro-bore system

29

HIGH TEMP. HOT WATER HEATING SYSTEM

Notes

1   The high temperature hot water heating system operates at a temperature of up to 200°C.

2   The system reduces the sizes of heat emitters and diameters of pipes.

3   Steam pressurisation is achieved by producing steam inside the boiler.

4   A mixing pipe is required with steam pressurisation to prevent water in the flow pipe from flashing into steam.

5   Nitrogen gas pressurises the water inside a pressure cylinder.

6   Nitrogen is an inert gas and if some of the gas is absorbed by the water it will not cause corrosion.

7   If the water in the cylinder, containing the nitrogen gas, falls too low, a pressure switch cuts in the boiler feed pump which forces water into the cylinder until the correct water level has been reached.

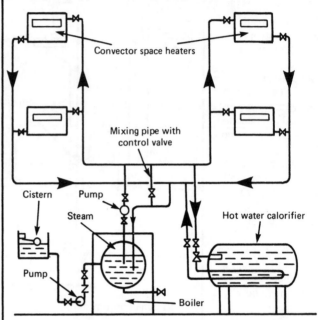

Fig 15 Steam pressurisation

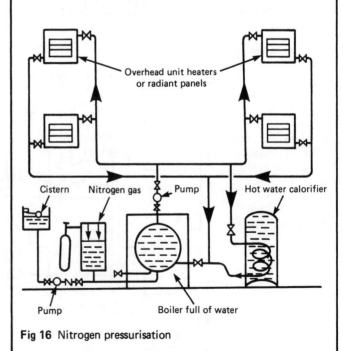

Fig 16 Nitrogen pressurisation

4 HEATING SYSTEMS – 2

STEAM HEATING SYSTEMS

Notes

1 Steam has a very high specific heat capacity of 2257 kJ/kg at atmospheric pressure, water has a specific heat capacity of 420 kJ/kg at 100°C.

2 This high heat capacity and the high velocity of flow of steam of up to 36 m/s allows very small pipes to be used.

3 Steam flows through the pipe work without the use of pumps which saves power.

4 Two types of systems are used which are known as gravity and mechanical.

5 The pressure of steam should be as low as practicable which increases the heat capacity.

6 High pressure steam may sometimes be required to overcome resistances on long runs of pipework with pressure reducing valves at various branches to buildings.

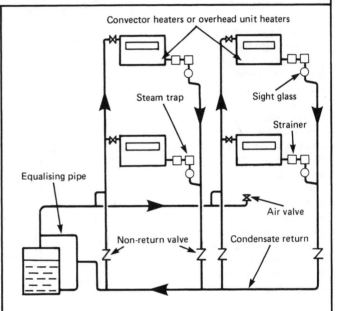

Fig 1 Gravity system

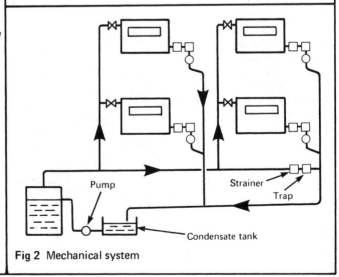

Fig 2 Mechanical system

STEAM TRAPS

Notes

1   The purpose of a steam trap is to allow condensate water to escape from the pipe work without allowing the steam to escape.

2   When water enters the thermostatic type the bellows contracts and opens the valve; steam closes the valve.

3   When water enters the bucket, the bucket sinks and the valve opens, steam forces the water out and the valve closes.

4   When water enters, the ball-float rises and the valve opens. The water is forced out by the steam and the valve is closed by the weight of the float and the steam pressure acting on the valve.

5   To prevent damage to the valves by grit, a strainer should be fitted to the inlet of the trap.

6   A sight glass fitted on the outlet of the trap will give the engineer a visual indication that water is passing through and not steam.

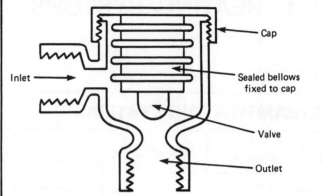

Fig 3 Thermostatic type

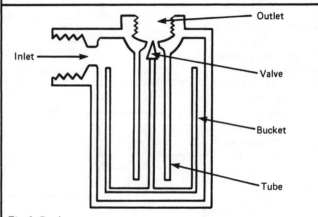

Fig 4 Bucket type

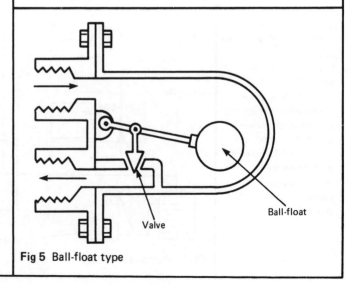

Fig 5 Ball-float type

PANEL HEATING

Notes

1   The system consists of 15 mm or 22 mm o.d. soft copper pipes embedded in the floor, ceiling or walls.

2   The system provides a high degree of thermal comfort and radiant heat is given off from the building fabric.

3   Thermostatic control provides the following surface temperatures of the fabric surfaces:
Floors 27°C
Ceilings 49°C
Walls 43°C

4   The pipes must be laid to falls to allow air to escape through an air valve.

5   Joints on the pipes must be made by capillary soldered fittings or bronze welding.

6   Before embedding the pipes the system must be hydraulically tested to a pressure of 1400 kPa for 24 hours.

7   The system avoids unsightly pipes and radiators. A good distribution of heat is also obtained.

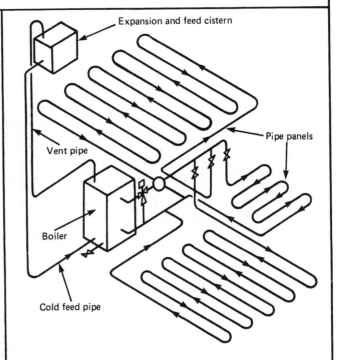

Fig 6 Installation of panel heating system

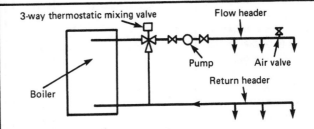

Fig 7 Detail of boiler and connections

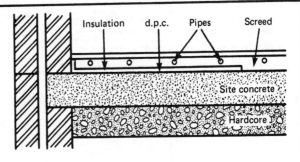

Fig 8 Method of embedding the panels

SOLAR SPACE HEATING

Notes

1   If solar space heating is to be installed thermal insulation of the building fabric is essential.

2   For a house having a floor area of 150 m² a flat plate solar collector having an area of 40 m² is required.

3   The solar tank should hold 40 m³ of water and should be provided with a valve which opens when the water temperature is too high and allows water to escape to waste.

4   The space heating may be either convector heaters or pipe panels embedded in the floors.

5   An anti-freeze must be added to the water in the solar circuit.

6   Because gas, oil, solid fuel and electricity keep rising in price, solar space heating is becoming more popular. During very cold weather a 'back up' heater is required, heated by one of the above fuels.

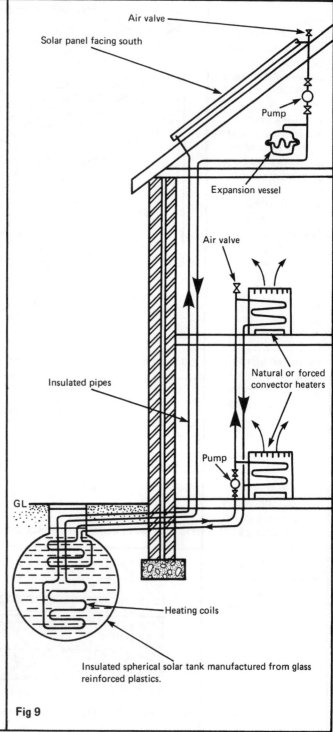

Air valve

Solar panel facing south

Pump

Expansion vessel

Air valve

Natural or forced convector heaters

Insulated pipes

Pump

GL

Heating coils

Insulated spherical solar tank manufactured from glass reinforced plastics.

Fig 9

WARM AIR HEATING SYSTEM

Notes

1　Warm air heating is cheap to install and may be used for full or partial heating.

2　The heater may be fired by gas, electricity, oil or solid fuel.

3　The system is easy to control and provides a quick warm-up period.

4　There is no risk of water leaks or freezing.

5　The air ducts must be well insulated.

6　Grilles are required in the doors so that air may return back to the heater.

7　An immersion heater or a gas circulator is required for the supply of hot water.

8　Air inlets are required for combustion air to the heater.

9　Fresh air is usually supplied to the rooms through openable windows but a fresh air duct may be connected to the return air duct.

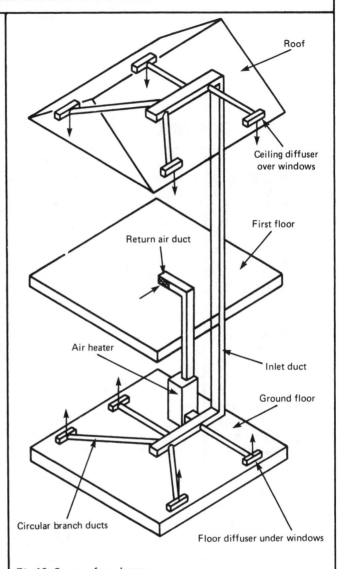

Fig 10 System for a house

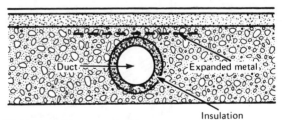

Fig 11 Duct inside concrete floor

5   HEATING SYSTEMS – 3

DISTRICT HEATING — 1

Notes

1   A district heating system is an extension of heating in one building to the heating of several buildings, or an entire town from one central boiler room.

2   The central boiler room should be sited close to buildings requiring a high heating load, e.g. a factory estate.

3   Long runs of heating pipes are required and these must be well insulated.

4   There is a saving in maintenance and waste heat from power stations may be used or refuse may be burned in the boilers.

5   There is a saving in boiler rooms and space.

6   There is less atmospheric pollution and only one chimney and fuel store.

7   The boilers and plant are maintained at peak efficiency.

8   Thermostatic control of the system is required.

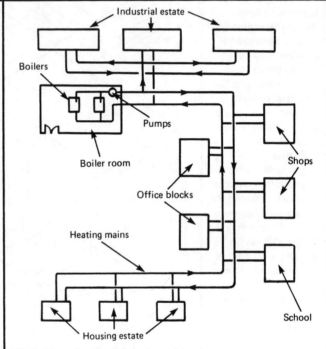

Fig 1 Plan of typical two-pipe scheme

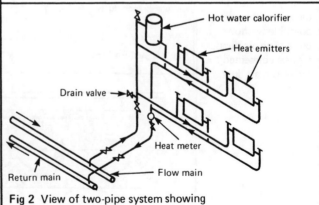

Fig 2 View of two-pipe system showing the internal distribution

DISTRICT HEATING — 2

Notes

1   The plan of the three-pipe system is similar to the plan of the two-pipe system except for the addition of a small diameter flow pipe connected to the boilers and laid alongside a large heating flow pipe.

2   The small diameter flow pipe is also provided with a separate pump.

3   When space heating is not required, the small flow pipe is used and the large flow-pipe shut off. This saves on heat losses on the pipes and power for the pumping.

4   The pipes must be laid at least 460 mm below ground and well insulated. The insulation must be kept dry and the pipes protected from corrosion.

5   The cost of the heating mains approximates to the cost of the boiler plant.

6   The heat losses from the heating mains is approximately 15% of the heating load to the heat emitters.

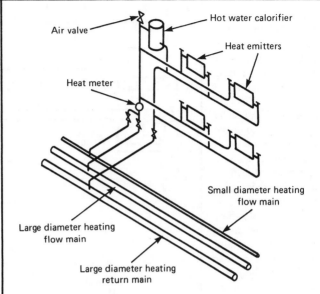

Fig 3 View of typical three-pipe system showing the internal distribution

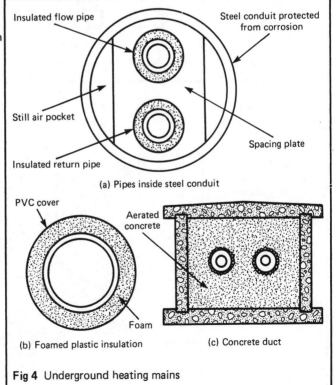

(a) Pipes inside steel conduit

(b) Foamed plastic insulation

(c) Concrete duct

Fig 4 Underground heating mains

DISTRICT HEATING — 3

Notes

1   The four-pipe system supplies both space heating and hot water to the buildings.

2   Hot water storage calorifiers are required in the boiler room and cold water storage cisterns will also be required to feed these calorifiers.

3   There is a great reduction in the number of calorifiers and cold water storage cisterns which would otherwise be required in each building.

4   The size of the boiler room would have to be increased to install the calorifiers.

5   During the period when space heating is not required only the hot water supply mains would be used.

6   The charge for space heating and hot water supply may be through the rents or rates of the properties served. Metering however is the fairest method of charging for heat.

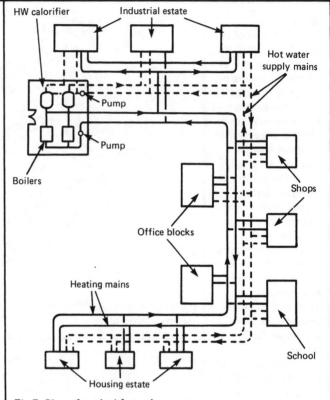

Fig 5 Plan of typical four-pipe system

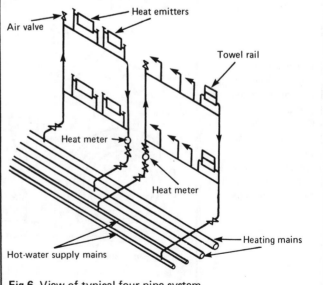

Fig 6 View of typical four-pipe system

38

HEAT EMITTERS — 1

Notes

1 In the absence of sufficient warmth from the sun or internal heat gains from lighting, people and machines, the heat losses from the body must be balanced by the installation of heat emitters.

2 The heat losses from the body approximate to the following:
Radiation 45%
Convection 30%
Evaporation 25%

3 A 3°C lower air temperature may be used when radiant heating is installed and this can save about 15% of fuel cost.

4 Aluminium foil at the back of radiators will improve their efficiency.

5 A shelf fitted over the radiator will prevent staining of the wall above the radiator due to convection currents.

6 Radiant panels and strips are often installed in factories and warehouses.

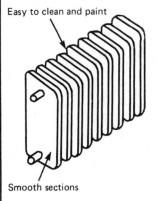

Easy to clean and paint

Smooth sections

Fig 7 Hospital type radiator

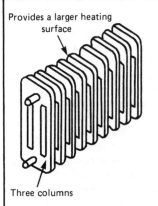

Provides a larger heating surface

Three columns

Fig 8 Column type radiator

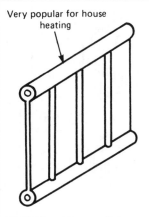

Very popular for house heating

Fig 9 Panel type radiator

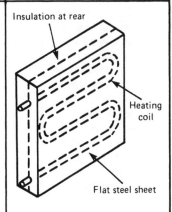

Insulation at rear

Heating coil

Flat steel sheet

Fig 10 Radiant panel

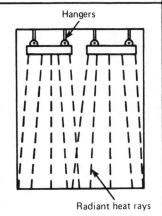

Hangers

Radiant heat rays

Fig 11 Radiant panels fixed overhead

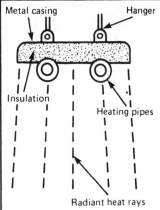

Metal casing   Hanger

Insulation

Heating pipes

Radiant heat rays

Fig 12 Radiant strip

HEAT EMITTERS — 2

Notes

1   The radiant and convector skirting heaters are very unobstrusive and provide a good distribution of heat in the room.

2   The natural convector must have the heater at low level so that the column warm air is displaced by the cooler air inside the room.

3   The fan convector may have the heater at high level. The fan may have variable speed to control the rate of heat output. The fan may also be used during summer to create air circulation.

4   The overhead unit heater is used in workshops and one may be used to form a warm air curtain across doorways.

5   The unit heater may have a thermo-statically-controlled inlet valve or several unit heaters may be controlled through a three-way diverting valve.

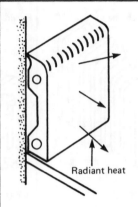

Fig 13 Radiant skirting heater

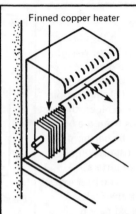

Fig 14 Convector skirting heater

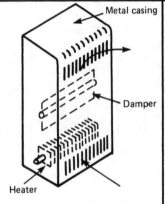

Fig 15 Natural convector

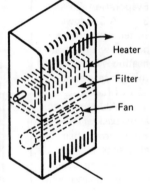

Fig 16 Fan convector

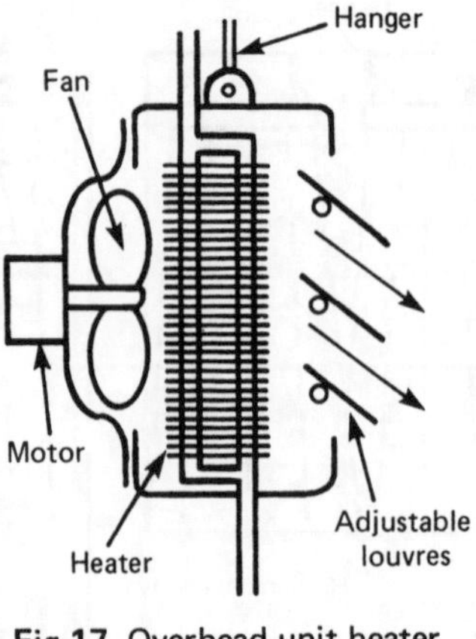

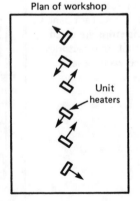

Fig 17 Overhead unit heater

Fig 18 Method of siting overhead unit heaters

THERMOSTATIC CONTROL OF HEATING SYSTEMS

Notes

1   Thermostatic control of heating systems saves fuel and improves the thermal comfort of the occupants.

2   The room thermostat should be sited away from draughts or direct sunlight and fixed between 1·2 m and 1·5 m above the floor level.

3   Thermostatic valves may be fixed to each heat emitter and this method will take into account the heat gains or losses in each room.

4   A cheaper method is by use of zoning valves to control the temperature of each circuit.

5   Three-port thermostatic valves may be either mixing or diverting.

6   The mixing valve has two inlets and one outlet and the diverting valve one inlet and two outlets.

7   Mixing and diverting valves prevent radiators from going cold and therefore provide a high degree of thermal comfort to the building occupants.

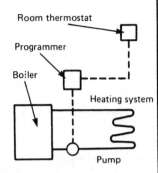

Fig 19 One thermostat controlling the pump

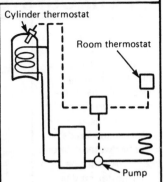

Fig 20 Two thermostats controlling the pump to give priority to hot water supply

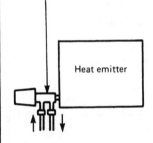

Fig 21 Thermostatic radiator valve

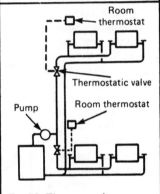

Fig 22 Thermostatic zoning valves

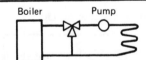

Fig 23 Mixing valve gives constant rate of flow and variable flow temperature

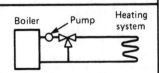

Fig 24 Diverting valve gives constant flow temperature and variable flow

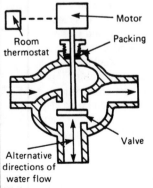

Fig 25 Section through a three port valve operated by a room thermostat

41

THERMOSTATIC AND CLOCK CONTROL OF HEATING SYSTEMS

Notes

1   The diverter valve may be used to close the heating circuit so that the water from the boiler can flow through the heating coil inside the hot water cylinder.

2   A rod-type thermostat may be used to control the temperature of the water in the cylinder. When the required water temperature is reached the brass casing expands and pulls the invar rod away from the contact points thus switching off the electrical supply.

3   A room thermostat also operates on the differential expansion of brass and invar.

4   The thermostatic radiator valve operates on a heat sensitive element which, when the correct air temperature is reached expands and closes the valve.

5   A clock controller and programmer will regulate the time at which the heating and hot water supply will operate.

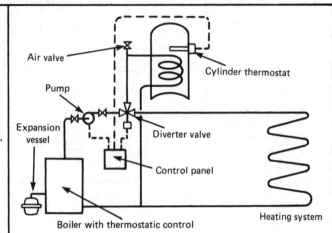

Air valve

Pump

Expansion vessel

Cylinder thermostat

Diverter valve

Control panel

Boiler with thermostatic control

Heating system

Fig 26 Use of diverter valve to give priority to hot water supply to a system having a pumped circuit to both the heating and the hot water cylinder

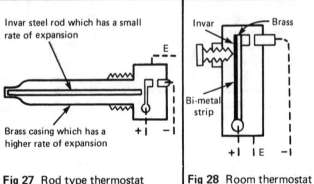

Invar steel rod which has a small rate of expansion

E

Brass casing which has a higher rate of expansion

Fig 27 Rod type thermostat

Invar   Brass

Bi-metal strip

Fig 28 Room thermostat

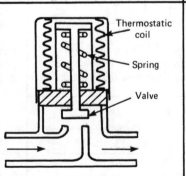

Thermostatic coil

Spring

Valve

Fig 29 Thermostatic radiator valve

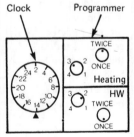

Clock   Programmer

TWICE
ONCE
Heating
HW
TWICE
ONCE

Fig 30 Clock control and programmer

42

6 STORAGE OF FUEL

Notes

1   When solid fuel or oil is to be used it is essential to consider the type of fuel storage.

2   For domestic or small buildings a brick or concrete bunker containing removeable timber boards may be used for solid fuel.

3   For larger buildings a fuel bunker or hopper above the boiler may be used for solid fuel.

4   The bunker or hopper allows the boiler to be fed automatically.

5   An oil storage tank may be sited either in the open air or in a room depending upon the local fire regulations.

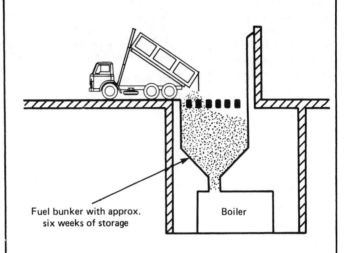

Fuel bunker with approx. six weeks of storage

Boiler

Fig 1 Solid fuel boiler in basement or sub-basement

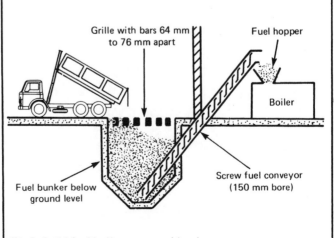

Grille with bars 64 mm to 76 mm apart

Fuel hopper

Boiler

Fuel bunker below ground level

Screw fuel conveyor (150 mm bore)

Fig 2 Solid fuel boiler at ground level

6   An oil tank room must be constructed of fire resisting material and the base of the room must be rendered with cement mortar so that the base will hold all the oil in the tank plus 10%.

7   Where the oil storage room is within the building it should be totally enclosed with walls and floors having not less than four-hours fire resistance.

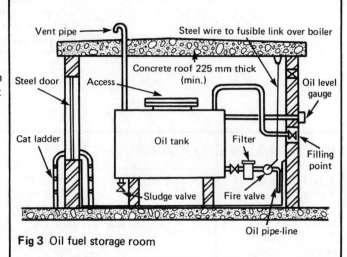

Fig 3 Oil fuel storage room

7 VENTILATION SYSTEMS

NATURAL VENTILATION — 1

Notes

1   Ventilation is the process of changing air in an enclosed space and its purpose is:
(a) to prevent depletion of oxygen content of the air;
(b) to prevent undue concentrations of carbon dioxide, body odours, tobacco smoke and moisture;
(c) to remove heat from lighting, people and machinery;
(d) to remove toxic gases and dust;
(e) to reduce bacteria.

2   Natural ventilation is cheap to install and maintain, does not use electrical power and is silent.

3   The forces necessary to create natural ventilation are
(a) wind pressure,
(b) stack pressure, i.e. convection or a combination of (a) and (b).

4   Unfortunately, if these natural forces are not present ventilation will cease and it may sometimes be necessary to use mechanical ventilation.

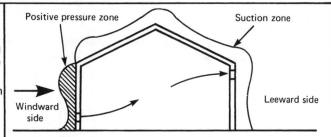

Fig 1 Wind pressure diagram for roofs with pitches up to 30°

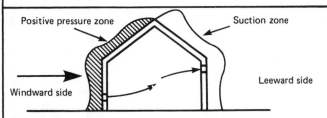

Fig 2 Wind pressure diagram for roofs with pitches above 30°

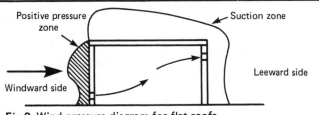

Fig 3 Wind pressure diagram for flat roofs

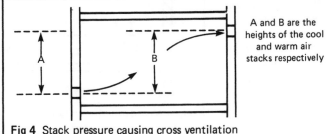

A and B are the heights of the cool and warm air stacks respectively

Fig 4 Stack pressure causing cross ventilation

NATURAL VENTILATION — 2

Notes

1   The rate of air change depends upon the type of building and the local regulations; public buildings usually require a ventilation rate of 28 m³ per person per hour.

2   Wind passing along the walls of a building produces a suction effect and air from rooms may be drawn out.

3   In tall buildings, during the winter months the cooler heavier outside air will tend to force the warmer lighter inside air out of the windows on the upper floors. This will produce draughts on the lower floors and excessive warmth on the upper floors.

4   Ventilation for an assembly hall or similar building may be produced by cool air from outside passing through a heat exchanger inside a convector and flowing out through high level ductwork. The flow of cold air through the heater must be controlled by a damper which is operated at the side of the heater.

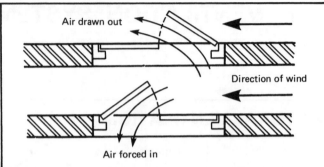

Fig 5 Wind causing ventilation through windows

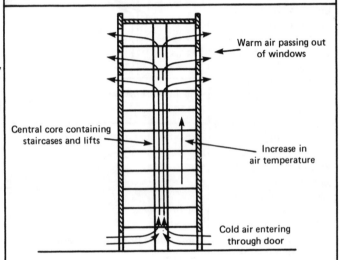

Fig 6 Stack pressure in a tall building

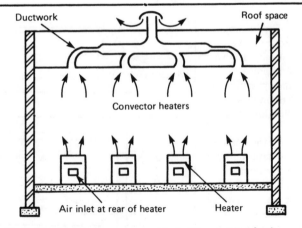

Fig 7 Ventilation for an assembly hall by passing fresh air through heat emitters

MECHANICAL VENTILATION — 1

Notes

1 Mechanical ventilation uses a fan or fans to create air change and movement.

2 There are three systems:
(a) natural inlet and mechanical extract;
(b) mechanical inlet and natural extract;
(c) mechanical inlet and mechanical extract.

3 The system, unlike natural ventilation, can be designed to provide a positive air change and air movement. However it costs more to install, operate and maintain. There is also a risk of noise from the fan and ducts.

4 Internal sanitary accommodation rooms must use a shunt duct to prevent smoke or smells passing from one room to another. Duplicated fans are also required with automatic changeover in the event of failure of the duty fan.

5 Basement car parks require an air change of about twenty per hour and duplicated fans with automatic changeover in the event of failure of the duty fan.

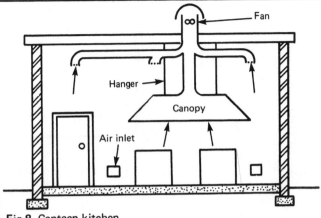

Fig 8 Canteen kitchen

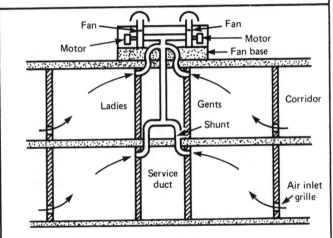

Fig 9 Internal sanitary accommodation

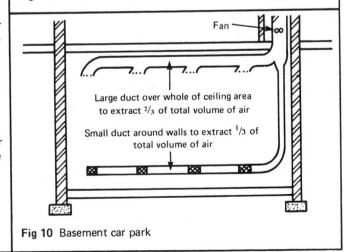

Large duct over whole of ceiling area to extract ²/₃ of total volume of air

Small duct around walls to extract ¹/₃ of total volume of air

Fig 10 Basement car park

MECHANICAL VENTILATION — 2

Notes

1   For habitable rooms the external air must be heated before it is forced into a room by a fan. It is also important that the amount of air extracted is carefully controlled or otherwise there will be a big loss of heat.

2   An additional extract duct and fan will provide a better form of control. The mechanical inlet and mechanical extract system is sometimes known as the balanced system of ventilation.

3   If air is extracted through the light fittings the heat from them can be circulated back to the heating unit. The efficiency of the lighting when ventilated, increases the light output by about 10%.

4   Where smoking is not permitted, as in a theatre, a downward air distribution system may be used. This system gives a good distribution of warm, filtered air.

5   The ductwork must be well insulated.

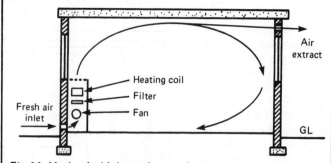

Fig 11 Mechanical inlet and natural extract

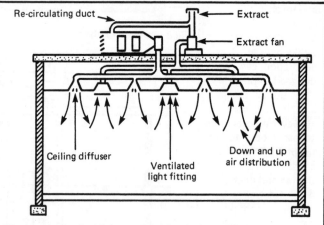

Fig 12 Mechanical inlet and mechanical extract for an open plan office or supermarket

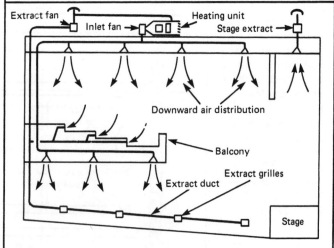

Fig 13 Mechanical inlet and mechanical extract for a theatre

TYPES OF FANS

Notes

1   The propeller fan does not create much air pressure and will not move air through long lengths of ductwork. The fan is for free air openings at windows and walls.

2   The axial flow fan can develop high pressure and is used for moving air along ductwork. The fan is fixed in the run of duct and does not require a base.

3   The bifurcated axial flow fan is used for moving hot gases e.g. flue gases.

4   The cross-flow fan is used inside fan convector units.

5   The centrifugal fan can develop high pressure for moving air through ductwork. It may have one or two inlets. A base is required for the fan. The fan has various forms of impeller depending upon the air condition.

6   There are three main laws governing the operation of a fan
(a) The discharge varies directly with the fan speed;
(b) The pressure developed varies with the square of the fan speed;
(c) The power absorbed varies with the cube of the fan speed.

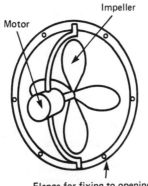

Fig 14 Propeller fan

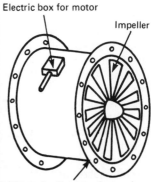

Fig 15 Axial flow fan

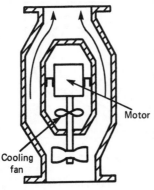

Fig 16 Bifurcated axial flow fan

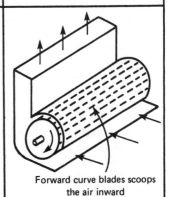

Fig 17 Cross-flow fan

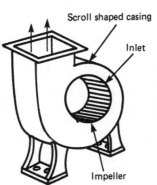

Fig 18 Centrifugal fan

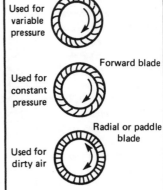

Fig 19 Types of impeller used with centrifugal fans

49

SOUND ATTENUATION IN DUCTWORK

Notes

1 Fans can be a major source of noise in air distributing systems. A fan revolving at high speed is usually noisier than when revolving at low speed.

2 Fans are not the only source of noise and ductwork, grilles, mixing boxes, elbows and bends may also generate noise. Large ducts may need to be stiffened to prevent drumming.

3 Fans may be mounted on concrete with either a cork or compressed glass fibre pad insert and the duct connected to the fan with a flexible connection. Alternatively the fan may be mounted on rubber or springs.

4 Sound attenuation of ducts may be made by use of perforated metal inserts or honeycomb inserts packed with acoustic material.

5 Bends and elbows may be provided with splitters which prevent turbulent flow of air.

6 Ducts passing through a high source of sound may be lined.

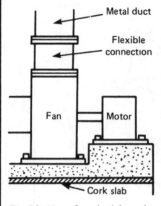

Fig 20 Use of cork slab and flexible connection

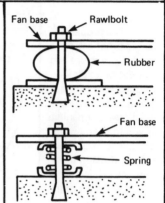

Fig 21 Use of rubber or spring mountings

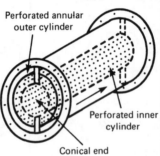

Fig 22 Use of perforated metal cylinder

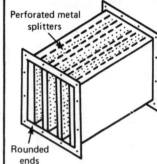

Fig 23 Use of perforated metal splitters

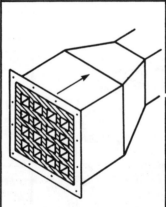

Fig 24 Use of acoustically absorbent honeycomb

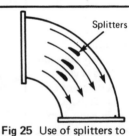

Fig 25 Use of splitters to give streamline flow

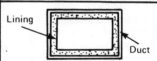

Fig 26 Use of acoustically absorbent lining of mineral wool

AIR FILTERS — 1

Notes

1   Cell filters can be arranged in a vee formation which increases the area of filter surface in contact with the air. The dry type can be vacuum cleaned but in time are thrown away. The viscous filter is coated with a odourless, non-toxic, non-flammable oil and can be cleaned in hot water and recoated with oil.

2   The bag type filter provides a large filter media in contact with the air to be cleaned.

3   The automatic roller filter is operated by a pressure sensitive switch on the inlet side. When the filter becomes dirty the air pressure is increased and the detector switches on the motor which brings clean fabric down from the top spool.

4   Several perforated rollers can be used which gives a vee formation and increases the area of filter in contact with the dirty air.

5   The efficiency of the filters range between 50 to 95%.

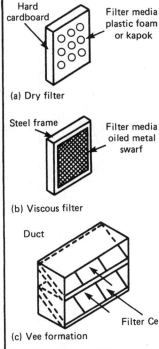

(a) Dry filter

(b) Viscous filter

(c) Vee formation

Fig 27 Cell-type filters

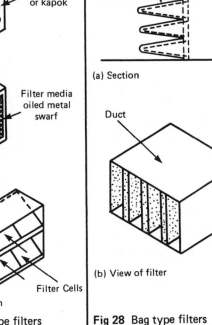

(a) Section

(b) View of filter

Fig 28 Bag type filters

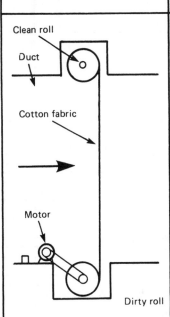

Fig 29 Automatic roller filter

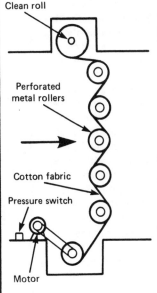

Fig 30 Automatic roller giving vee formation

AIR FILTERS — 2

Notes

1   Viscous filters have a high dust holding capacity and are often used in industrial areas.

2   An oil spray type has closely spaced corrugated metal plates, which are continuously coated and washed from a spray pipe.

3   A rotating type consists of filter plates hung from chains. The bottom plates pass through a bath of oil which cleans and re-coats them with clean oil.

4   An electrostatic filter has an ionising section which gives the dust particles a positive electrostatic charge. These positively charged particles then pass through plates which are positively charged and arranged so that another plate, which is earthed, is next to the positively charged plate. The positively charged dust particles are repelled by the positive plates and attracted to the negative plates. If required, an activated carbon filter may be fitted after the electrostatic filter. The efficiency of the filter is about 99%.

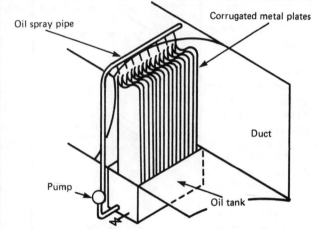

Fig 31 Automatic viscous filter (oil-spray type)

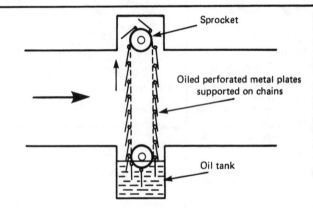

Fig 32 Automatic viscous filter (rotating type)

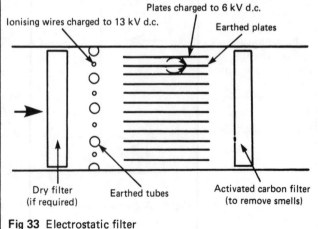

Fig 33 Electrostatic filter

52

8 AIR CONDITIONING

AIR CONDITIONING — 1
(Psychrometric process)

Notes

1 A psychrometric chart concerns the behaviour of mixtures of air and water vapour.

2 By sensible heating or cooling of air the relative humidity may either be too low or too high respectively. Figs 3 and 4 show this process.

3 If the air enters the plant at 5°C and with a RH of 60% and the room air is required at 20°C with a RH of 50% the air will require preheating to 18·5°C, cooled to 9°C dewpoint temperature and then finally heated to 20°C. (See Fig 5).

4 If the air enters the plant at 30°C and with a RH of 70% and the room air is required at 20°C with a RH of 50% the air will require cooling to 9°C dewpoint temperature and then re-heated to the required temp. of 20°C (See Fig 6).

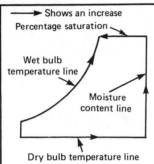

Fig 1 Use of psychrometric chart

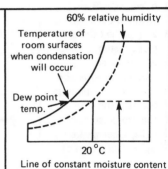

Fig 2 Condensation on room surfaces

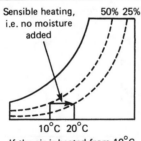

Fig 3 Heating of air without adding moisture

If the air is heated from 10°C to 20°C the RH = 25%

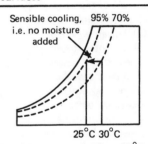

Fig 4 Cooling of air without dehumidification

If the air is cooled from 30°C to 25°C the RH = 95%

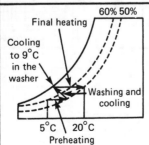

Fig 5 Humidifying by preheating, washing and final heating

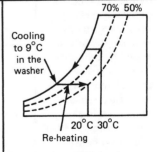

Fig 6 Dehumidifying by cooling, washing and re-heating

AIR CONDITIONING — 2
(Central plant system)

Notes

1   An air conditioning system controls within predetermined limits the temperature, relative humidity, cleanliness and movement or air within a building.

2   The central plant system is used where the air temperature may be the same throughout the building e.g. theatres, assembly halls, supermarkets.

3   Operation of main unit:

(a) Fresh air enters and is mixed with up to 75% of the re-circulated air.

(b) The air is filtered.

(c) In winter the air is heated and passes through the washer where it is cooled to dew point temp.

(d) In summer the air is also cooled to dew point temp. in the washer which causes dehumidification.

(e) The air passes through zig-zag eliminator plates which remove drops of water and any dirt missed by the filter.

(f) The air passes through the final or reheater which adjusts the final temperature and final relative humidity.

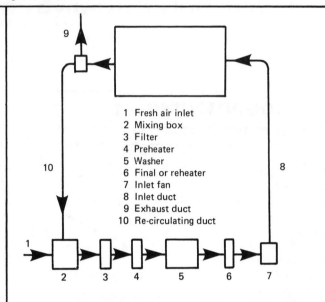

1 Fresh air inlet
2 Mixing box
3 Filter
4 Preheater
5 Washer
6 Final or reheater
7 Inlet fan
8 Inlet duct
9 Exhaust duct
10 Re-circulating duct

Fig 7 Diagrammatical layout of central plant only system

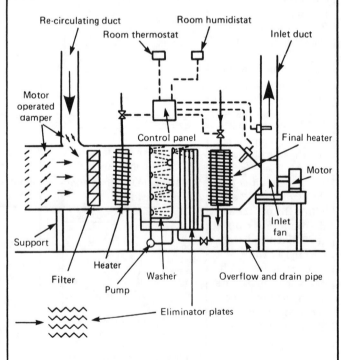

Fig 8 Section of main unit for the central plant system

AIR CONDITIONING — 3
(Variable air volume)

Notes

1 The variable volume system consists of a central station operating at a set temperature and relative humidity which varies with the outside conditions.

2 The conditioned air from the main unit passes to outlets, usually in the ceiling, which are provided with thermostatically controlled actuators.

3 In a large room several of these variable air volume ceiling units may be controlled by one room thermostat.

4 Several rooms may have separate thermostats to control the flow of air into each room.

5 The inlet fan may have variable pitched blades operated by compressed air. A pressure switch controls the pitch of these blades.

6 The air distribution system is usually either medium or high velocity.

7 The temperature of the air in each zone can be varied but the system is only suitable for buildings having fairly evenly distributed cooling load.

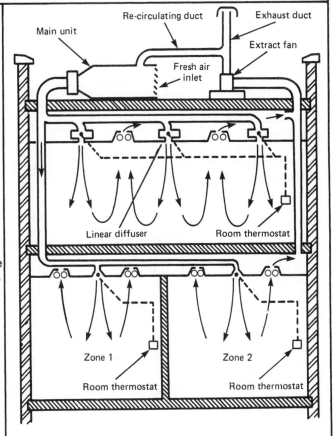

Fig 9 Layout of a typical variable air volume system

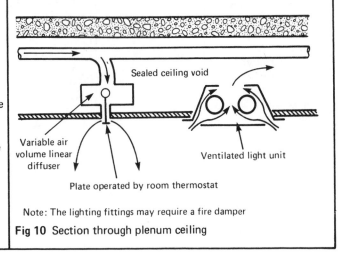

Note: The lighting fittings may require a fire damper

Fig 10 Section through plenum ceiling

55

AIR CONDITIONING — 4
(Induction system)

Notes

1   As the name implies, the air in the rooms is induced into the unit by means of nozzles.

2   The primary conditioned air is forced into the units at high velocity and this air is mixed with the secondary room air.

3   A damper controls whether or not the room air passes through a heating coil which is connected to a room thermostat.

4   If required, the heating coils, in summer, may be changed over to cooling coils.

5   If heating only is used the system is known as the two-pipe induction system and when four pipes are used (two for heating and two for cooling) the system is known as the four-pipe "change-over" induction system.

6   The latter system gives excellent control of the air temperature in various zones.

7   If the induction unit is fitted below a window, warm air prevents condensation in winter and cools the window in summer.

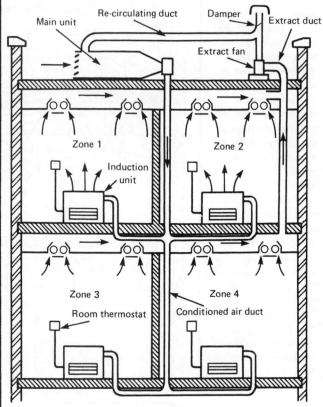

Fig 11 Layout of typical induction system

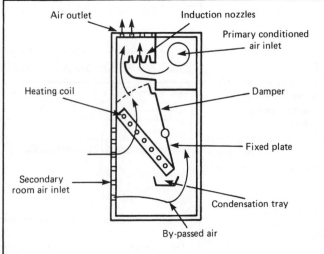

Fig 12 Section through an induction room unit

AIR CONDITIONING — 5
(Fan-coil unit and air washer)

Notes

1   Instead of the secondary room air being drawn through the terminal unit by nozzles, as in the case of the induction system, outlined on the previous page, a fan-coil terminal unit may be used.

2   When the fan coil unit is used the method is known as the fan-coil air conditioning system. Its disadvantage is the additional cost of the electric wiring and fans. Power for the fans will also be required. The layout of the system is similar to the induction system.

3   The main function of the air washer is for humidifying or dehumidifying the air but it also removes some of the dirt. The water pressure for the spray is usually between 200 and 300 kPa and the velocity of the air through the washer is usually between 2 and 2.5 m/s. The tank will require periodic cleaning.

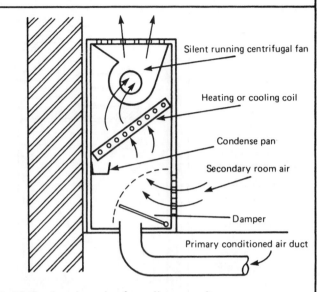

Silent running centrifugal fan

Heating or cooling coil

Condense pan

Secondary room air

Damper

Primary conditioned air duct

Fig 13 Section through a fan-coil room unit

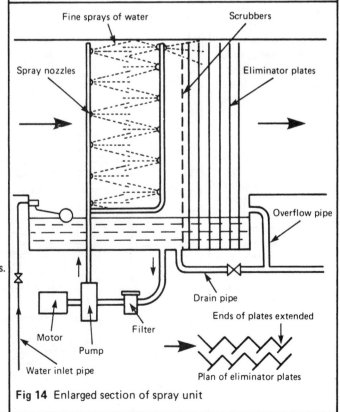

Fine sprays of water

Scrubbers

Spray nozzles

Eliminator plates

Overflow pipe

Drain pipe

Ends of plates extended

Motor

Filter

Pump

Water inlet pipe

Plan of eliminator plates

Fig 14 Enlarged section of spray unit

AIR CONDITIONING — 6
(Dual duct system)

Notes

1   The dual duct system is, like the variable volume and induction systems, used for rooms requiring varying air temperature.

2   Water circulation in the system is eliminated and the system operates on an 'all year round' principle, in other words the rooms can be both cooled or warmed in both winter and summer.

3   The cool air duct operates at about 8°C in summer and about 16°C in winter. The hot air duct operates at about 20°C in summer and about 38°C in winter.

4   The heated and cooled air pass at high velocity to room mixing units.

5   The volumes of hot and cold air are controlled by a damper connected to a room thermostat and a constant volume plate is also incorporated.

6   Fire dampers are required inside the ducts where they pass through the floors. Fire dampers operated by smoke detectors, are also required at the outlets of the light fittings.

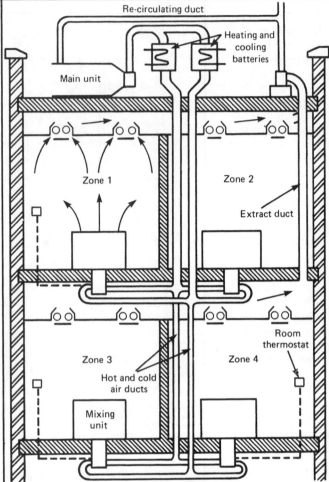

Fig 15 Layout of a typical dual-duct system

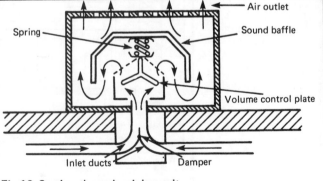

Fig 16 Section through mixing unit

HEAT PUMP — 1

Notes

1 The heat pump is a device which extracts heat from a low-temperature source and upgrades it to a higher temperature so that it may be used for space or water heating.

2 The low temperature heat source, may be from water, air or soil which surrounds the evaporator.

3 The heat pump will always give out more energy than the energy used for driving it and it is a means of using electrical energy to its best advantage.

4 The theoretical coefficient of performance (COP) is expressed as:

$$COP = \frac{t_c}{t_c - t_e}$$

where:-

t_c = condenser temp. in degrees Kelvin

t_e = evaporator temp. in degrees Kelvin

5 In very cold weather it is however necessary to use a separate boost heater to supplement the heat pump.

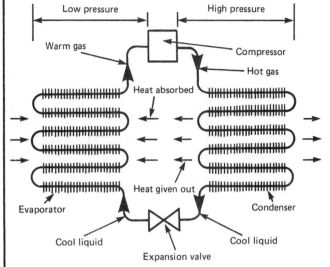

Note:- The flow of the refrigerant can be reversed so that the building is warmed in winter and cooled in summer

Fig 17 Principles of operation of the heat pump

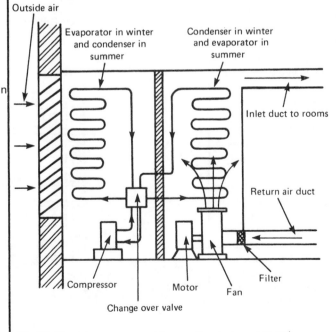

Fig 18 The heat pump used for cooling in summer and warming in winter

HEAT PUMP — 2

Notes

1   Besides large heat pump units that can be used to warm the whole building, small self-contained units are available which are usually fitted under each window which will create a warm air curtain in winter and a cool air curtain in summer.

2   In order to extract warm air from an extract duct, water may be circulated by a pump through a coil in the warm air extract duct to another coil in the cold air inlet duct. The efficiency of this method of heat recovery will be increased by use of a heat pump with an evaporator coil in the warm extract duct and a condenser coil in the cold air inlet duct.

3   Heat contained in warm waste water from baths, showers, basins and sinks may be used to warm a house by use of a heat pump. A large insulated tank is buried below ground to receive the waste water and heat from the tank extracted through an evaporator inside the tank.

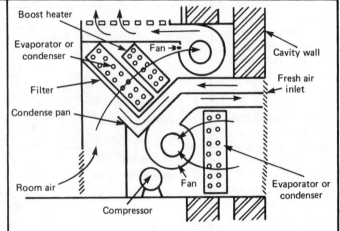

Fig 19 Unit heat pump fixed below window

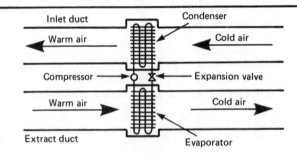

Fig 20 Heat pump used for heat recovery

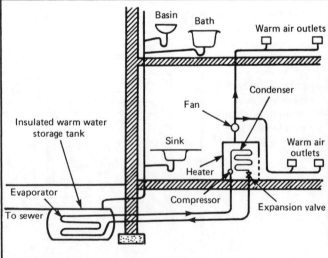

Fig 21 Heat pump used for extracting heat from warm waste water

HEAT RECOVERY DEVICES

Notes

1 The thermal wheel was invented by Carl Munter a Swedish engineer and is sometimes called the Munters' wheel. The wheel is from 600 mm to 3·6 m in diameter and is filled with wire mesh 1 mm to 1·5 mm across.

It is driven by an electric motor having a power of 700 W (max) and revolves at an angular velocity of 10 rev/min. The heat from the exhaust air is transferred to the inlet air and the purging section extracts the contaminants. The efficiency is up to 90%.

2 In the heat recovery duct the exhaust warm air is separated from the inlet cool air by metal or glass vanes. Heat from the exhaust vanes is transferred to the inlet vanes to heat the inlet air.

3 The ducts must be insulated to conserve heat and prevent condensation.

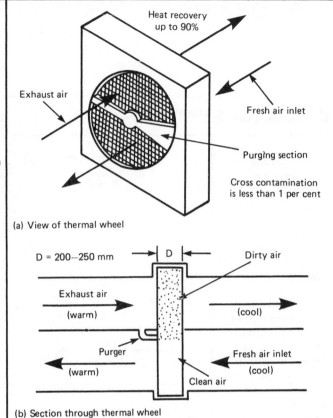

(a) View of thermal wheel

Heat recovery up to 90%

Exhaust air

Fresh air inlet

Purging section

Cross contamination is less than 1 per cent

D = 200–250 mm

Dirty air

Exhaust air (warm)

(cool)

Purger

Fresh air inlet

(warm)

(cool)

Clean air

(b) Section through thermal wheel

Fig 22 The thermal wheel

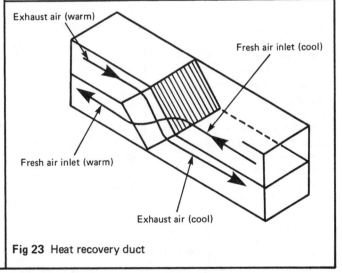

Exhaust air (warm)

Fresh air inlet (cool)

Fresh air inlet (warm)

Exhaust air (cool)

Fig 23 Heat recovery duct

9 DRAINAGE SYSTEMS

DRAINAGE SYSTEMS — 1

Notes
Key to symbols
IC = Inspection
chamber
RWG = Rainwater
gully.
WG = Waste gully.
RG = Road gully.
S & VP = Soil and vent
pipe.
YG = Yard gully.
RWS = Rainwater shoe.
RP = Rodding point.
1   In the combined
system of drainage,
the foul water from
the sanitary appliances
and the rainwater
from roofs and other
surfaces are carried by
a single drain to a
combined sewer. The
system is cheap but
the load on the sewage
works is high.
2   In Fig 2, the
foul water from the
sanitary appliances is
carried by a foul water
drain to a foul water
sewer and the rain-
water from the roofs
and other surfaces is
carried by a surface
water drain into a
surface water sewer.
The system is more
expensive but the
load on the sewage
works is greatly
reduced.

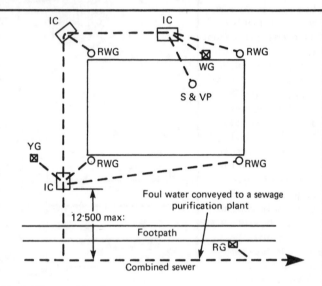

Fig 1 The combined system

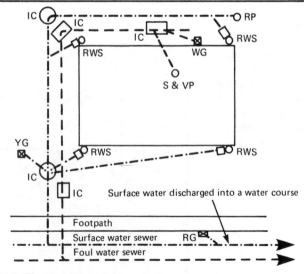

Fig 2 The separate system

DRAINAGE SYSTEMS — 2

Notes

1   In the partially separate system most of the rainwater is carried by the surface water drain into the surface water sewer. In order to save cost however, an isolated rainwater inlet is connected to the foul water drain. In Fig 3 the rainwater inlet at A is connected to the foul water drain thus saving a length of drain and also a rodding point at B replaces an inspection chamber.

2   A back inlet gully can be used for connecting the rainwater pipe or waste pipe to the drain.

3   The yard gully is similar to a road gully but is smaller.

4   The rainwater shoe is used for connecting a rainwater pipe to a surface water drain.

5   The soil and vent pipe is connected to the foul water drain by means of a rest bend. A concrete block should be inserted under the rest to provide additional support.

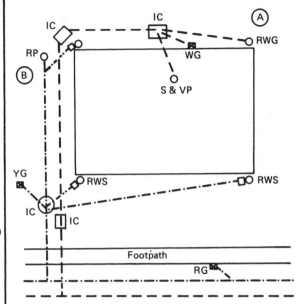

Fig 3 The partially separate system

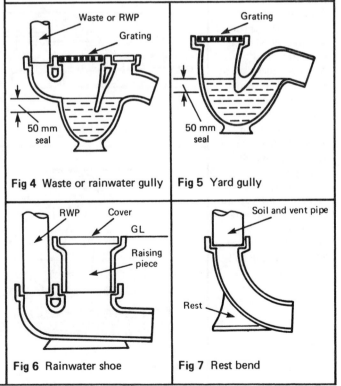

Fig 4 Waste or rainwater gully

Fig 5 Yard gully

Fig 6 Rainwater shoe

Fig 7 Rest bend

THE RODDING POINT SYSTEM

Notes

A report on the possible causes of blockages in drainage systems mentioned that manholes were sometimes one of the causes. Faulty benching connections are sometimes found.

1   The rodding point system is a much cheaper method of gaining access for rodding than by use of manholes and there is virtually no extra load on the draining system and therefore less risk of settlement. The system is also neater in appearance. uPVC pipes are used bedded and surrounded in granular material. Special flexible rods are used for clearing a blockage.

2   For drainage at the top of the run a shallow rodding point is used.

3   Deep drain runs will require a rodding point with a long radius bend.

4   Before the rodding point system is used it is essential to consult the Local Authority to ensure that the system will be permitted.

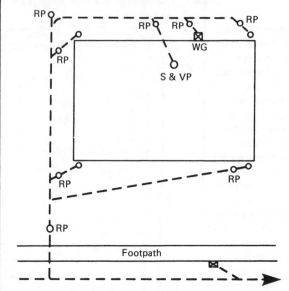

Fig 8 Plan of rodding point system

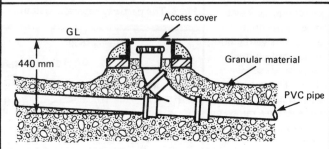

Fig 9 Shallow rodding point

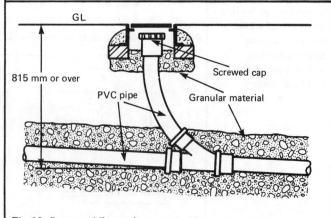

Fig 10 Deep rodding point

CONNECTION OF DRAINAGE TO SEWER

Notes

The connection of a drain to another drain or sewer, or a private sewer to a public sewer must be made obliquely in the direction of flow.

1   Drains from buildings may be connected separately to the public sewer so that each owner of the separate buildings will be responsible for the maintenance of the drainage system for that building.

2   In order to save on the cost of the separate connections to the public sewer a private sewer may be used, which will result in only one connection for several buildings. The maintenance of the private sewer will be shared between the separate owners.

3   The connection of drain to the public or private sewer can be made either by a junction inserted into the sewer or by means of a saddle. A hole is made in the top half of the sewer and the saddle is then inserted, bedded in cement mortar and surrounded in concrete.

4   Alternatively, the connection of the drain to the sewer may be made by inserting a junction.

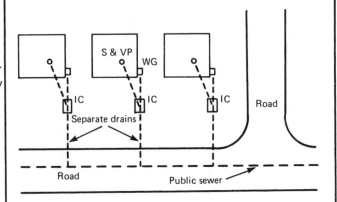

Fig 11 Use of separate drains

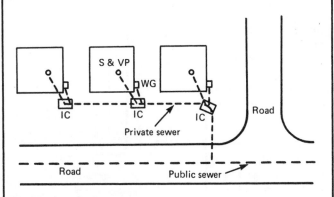

Fig 12 Use of private sewer

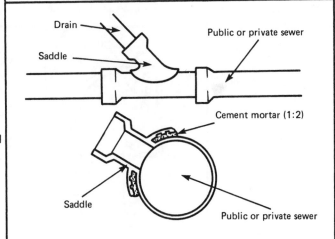

Fig 13 Use of saddle connection

DRAINAGE VENTILATION

Notes
The venting of foul water drains removes gases and ensures that the air inside the drain remains at atmospheric pressure which will prevent the loss of gully trap water seals by siphonage or compression.

1   The usual method of venting is by a direct connection to the public sewer without the use of an interceptor trap. This method is cheaper, there is less risk of blockage and the sewer is also well vented.

2   For old sewers, or for an area in which interceptors are already in use it is advisable to use the trap on the new drainage system.

3   In order to permit fresh air into the drainage system, a fresh air inlet is required which will prevent foul air from escaping.

4   The interceptor trap is provided with a water seal 64 mm deep and a rodding arm. The access stopper is provided with a chain which is fixed to a hook at the top of the manhole.

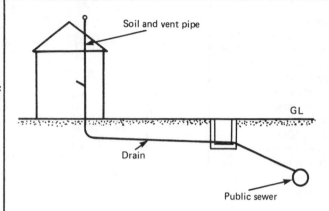

Fig 14 Without the use of an interceptor trap

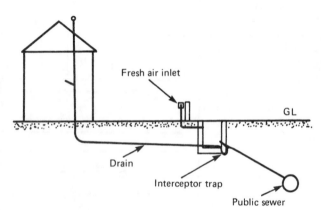

Fig 15 With the use of an interceptor trap

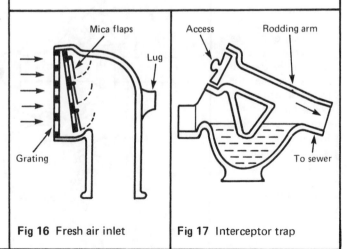

Fig 16 Fresh air inlet

Fig 17 Interceptor trap

DRAIN LAYING — 1

Notes

1   The bottom of the drain trench must be prepared with the required fall or gradient.

2   Sight rails are set up, painted black and white to mark the centre of the drain.

3   These sight rails are fixed above the trench and are given the gradient required for the drain. At least three sight rails must be used.

4   By means of a boning rod, sighted between the sight rails, the level of the trench bottom, with the correct gradient is obtained.

5   Wooden pegs are driven into the trench bottom level with the bottom of the boning rod and the soil dug out to the top of the pegs. The pegs are removed before drain laying is commenced.

6   For the safety of working in the trench it is essential to provide planking and strutting to the soil.

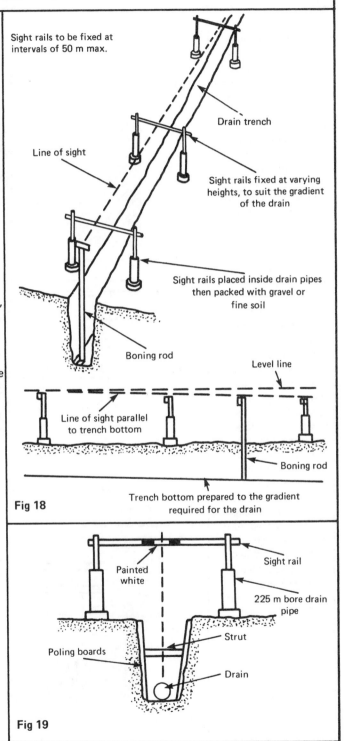

Sight rails to be fixed at intervals of 50 m max.

Drain trench

Line of sight

Sight rails fixed at varying heights, to suit the gradient of the drain

Sight rails placed inside drain pipes then packed with gravel or fine soil

Boning rod

Level line

Line of sight parallel to trench bottom

Boning rod

Fig 18

Trench bottom prepared to the gradient required for the drain

Sight rail

Painted white

225 m bore drain pipe

Poling boards

Strut

Drain

Fig 19

INSPECTION CHAMBERS

Notes

Inspection chambers are comparatively shallow. They are located at points where drain blocking is most likely to occur.

1   The traditional inspection chamber is built of brickwork either half brick or one brick in thickness depending on its depth below ground. The chamber should be built in English bond with flush internal joints. Inspection chambers must not be rendered internally as this can flake off and may block the drain.

2   Precast concrete chambers are easier and quicker to build. In wet ground the chamber should be surrounded in concrete 150 mm thick.

3   Unplasticised polyvinyl chloride (uPVC) chambers may be obtained having two branch connections on either side and with depths up to 540 mm and 940 mm. They are light in weight and quick to build.

3   The step irons should be fixed at 300 mm centre both in the vertical and horizontal plane.

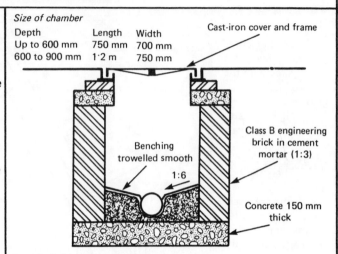

Size of chamber

Depth	Length	Width
Up to 600 mm	750 mm	700 mm
600 to 900 mm	1·2 m	750 mm

Cast-iron cover and frame

Benching trowelled smooth

1:6

Class B engineering brick in cement mortar (1:3)

Concrete 150 mm thick

Fig 20 Brick inspection chamber

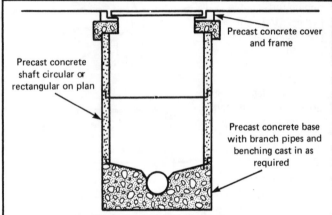

Precast concrete cover and frame

Precast concrete shaft circular or rectangular on plan

Precast concrete base with branch pipes and benching cast in as required

Fig 21 Precast concrete inspection chamber

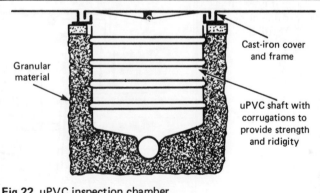

Cast-iron cover and frame

Granular material

uPVC shaft with corrugations to provide strength and ridigity

Fig 22 uPVC inspection chamber

THE MARSCAR BOWL — POSITIONS OF INSPECTION CHAMBERS

Notes

1 The Marscar access bowl is an alternative method to the inspection chamber and rodding point methods of access to drainage. Being near the surface it reduces costs and makes access easier. The branch pipe connections to the bowl are made by first cutting a hole, and then inserting the branch pipe, and making the joint between the pipe and the bowl by means of a solvent cement.

2 An inspection chamber is required at every change of direction either in the horizontal or vertical run of drain.

3 An inspection chamber is required at a junction or as close as possible to the junction.

4 On a straight run of drain or private sewer an inspection chamber is required at intervals of not more than 90 m.

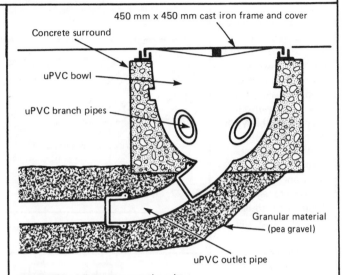

450 mm x 450 mm cast iron frame and cover
Concrete surround
uPVC bowl
uPVC branch pipes
Granular material (pea gravel)
uPVC outlet pipe

Fig 23 The Marscar access bowl

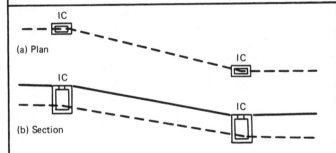

IC

(a) Plan

IC

IC

(b) Section

IC

Fig 24 Inspection chambers at change of direction

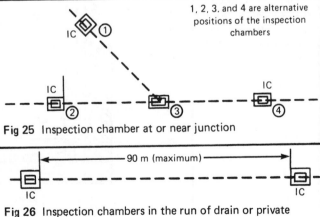

1, 2, 3, and 4 are alternative positions of the inspection chambers

IC ①

IC ②

③

IC ④

Fig 25 Inspection chamber at or near junction

90 m (maximum)

IC

IC

Fig 26 Inspection chambers in the run of drain or private sewer

BACK-DROPS

Notes

1   Where there is a great difference in level between the drain and the private or public sewer a back-drop will save a great deal of excavation which should pay for the cost of the manhole and back-drop.

At one time back drops were used on sloping sites to limit the gradient of the drain. This is unnecessary and the drain may be laid at the same slope of the surface of the ground.

2   For pipes up to 150 mm bore the back-drop may be fixed inside the manhole in cast-iron uPVC or pitch-fibre pipe.

For back-drops above 150 mm bore the back-drop is fixed outside the manhole and surrounded in concrete. The access shaft should be 750 mm min. and the working area 1·2 m x 750 mm min.

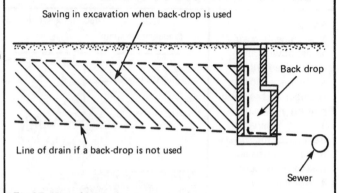

Saving in excavation when back-drop is used

Back drop

Line of drain if a back-drop is not used

Sewer

Fig 27 Use of back-drop

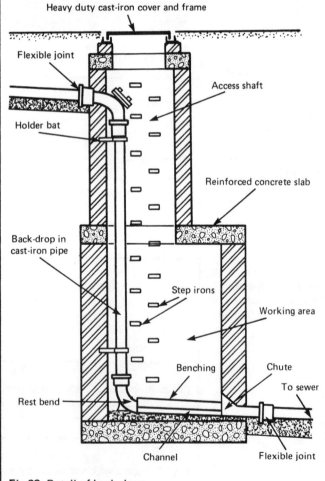

Heavy duty cast-iron cover and frame

Flexible joint

Access shaft

Holder bat

Reinforced concrete slab

Back-drop in cast-iron pipe

Step irons

Working area

Benching

Chute

To sewer

Rest bend

Channel

Flexible joint

Fig 28 Detail of back-drop

BEDDING OF DRAINS — 1

Notes

The term bedding factor means the ratio of the pipe strength when bedded, to the pipe strength given in the relevant British Standard.

1   Class A bedding gives a bedding factor of 2·6 which means that when bedded, the load that can be placed on top of the pipe may be 2·6 greater than the quoted BS pipe strength. This is due to the cradling effect of the concrete. The bedding is used where extra pipe strength is necessary or where greater accuracy in pipe gradient is required.

2   Class B bedding is very popular and is much cheaper and quicker to use than Class A. If it is used for plastic pipe it is essential to both bed and surround the pipe with granular material to prevent the pipe from being distorted.

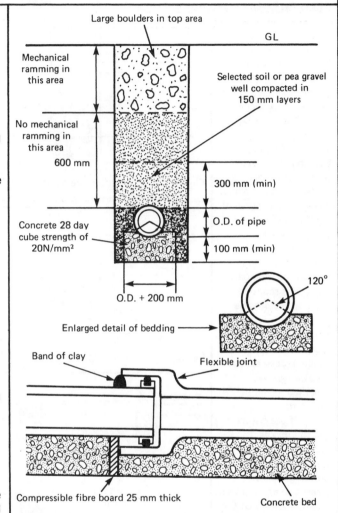

Large boulders in top area

GL

Mechanical ramming in this area

Selected soil or pea gravel well compacted in 150 mm layers

No mechanical ramming in this area

600 mm

300 mm (min)

Concrete 28 day cube strength of 20N/mm²

O.D. of pipe

100 mm (min)

O.D. + 200 mm

120°

Enlarged detail of bedding

Band of clay

Flexible joint

Compressible fibre board 25 mm thick

Concrete bed

Fig 29 Class A bedding: bedding factor 2·6

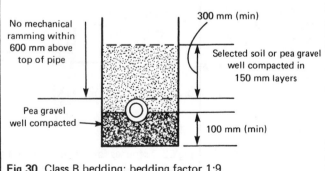

No mechanical ramming within 600 mm above top of pipe

300 mm (min)

Selected soil or pea gravel well compacted in 150 mm layers

Pea gravel well compacted

100 mm (min)

Fig 30 Class B bedding: bedding factor 1·9

BEDDING OF DRAINS — 2
(Drains near foundations)

Notes

If the soil is free from hard objects the cost of excavation and material can be reduced by the use of either Class C or Class D bedding.

1   In Class C bedding the trench bottom is shaped to receive the pipe which gives a small degree of craddling effect.

2   In Class D bedding the trench bottom is levelled to receive the pipe barrel and soil is scooped away to receive the pipe sockets. This method does not increase the load that can be placed on the pipe above that specified in the British Standard.

3   If the nature of the soil makes it necessary, a drain trench close to a building must be treated so as to prevent loss of stability of the foundation. The concrete fill must be provided with expansion joints at intervals of 9 m. Compressible fibre boards in between each length of concrete can be used.

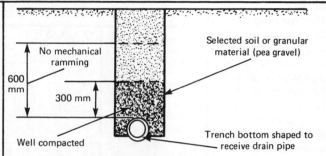

Fig 31 Class C bedding: bedding factor 1·1

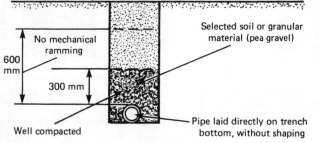

Fig 32 Class D bedding: bedding factor 1·0

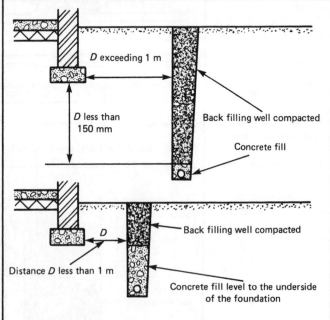

Fig 33 Trenches for drains or private sewers adjacent to foundations. Building Regulations N 14.

JOINTS USED ON DRAIN PIPES

Notes

1 Whenever possible flexible joints should be used which have the following advantages over rigid joints:

(a) They are quicker to make.

(b) The pipeline can be tested immediately.

(c) They resist ground movement without fracturing.

(d) No delay in jointing due to wet or freezing weather.

2 Clay pipes can be jointed with cement and sand mortar or one of several types of flexible joint. One of the most popular flexible joint is the Hepsleve.

3 Cast iron pipes can be jointed by caulked lead which must be cold before caulking. A recently introduced flexible joint on cast-iron pipes consists of a rubber sleeve held over the joint by two cast iron couplings bolted together.

4 uPVC pipes may be jointed by a solvent cement or by a rubber 'O' ring.

5 Pitch fibre pipes are jointed by a snap rubber 'D' ring.

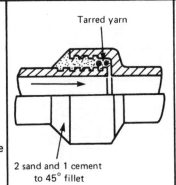

2 sand and 1 cement to 45° fillet

Fig 34 Cement mortar joint on clay pipe

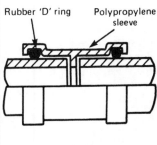

Pipe is lubricated and pushed into the sleeve

Fig 35 Hepsleve flexible joint on clay pipe

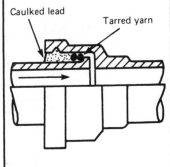

Fig 36 Caulked lead joint on cast-iron pipe

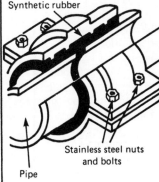

Fig 37 Flexible joint on cast-iron pipe

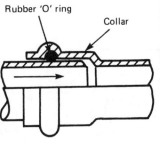

Pipe is lubricated and pushed into collar

Fig 38 Flexible joint on uPVC pipe

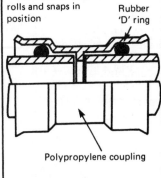

Fig 39 Flexible joint on pitch fibre pipe

ANTI-FLOOD DEVICES — GREASE TRAP

Notes

If there is a risk of the drain being flooded due to surcharge of the sewer some form of anti-flooding device will be required.

1   For drainage systems without an interceptor trap an anti-flooding trunk valve may be fitted in the drain run, inside the manhole nearest the sewer.

2   If an interceptor trap is to be used an anti-flooding type can be used instead of the conventional interceptor type.

3   If back flooding occurs through a gully trap an anti-flooding type can be used.

4   Waste water from canteen sinks or dish washers contains a considerable amount of grease which if not removed would cause blocking of the drain. In the grease trap the grease is cooled by a large volume of water. The grease solidifies and floats to the surface of the water. At regular intervals the tray is lifted out of the trap which also removes the grease.

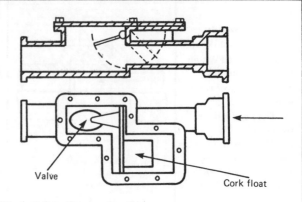

Fig 40 Anti-flooding trunk valve

Valve

Cork float

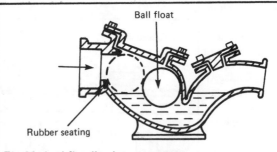

Ball float

Rubber seating

Fig 41 Anti-flooding interceptor trap

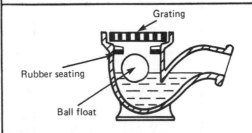

Grating

Rubber seating

Ball float

Fig 42 Anti-flooding gully trap

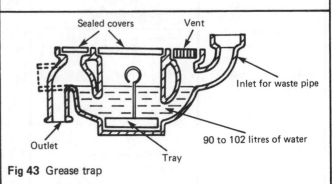

Sealed covers   Vent

Inlet for waste pipe

Outlet

Tray

90 to 102 litres of water

Fig 43 Grease trap

GARAGE DRAINAGE

Notes

The Public Health Act 1936 prohibits the discharge of petrol and oil into a sewer.

1   The floor washings of large garages must be prevented from allowing petrol or oil into the sewer. The floor should be arranged so that one garage fully takes the washings of up to 50 m² of floor area. The gully will retain some of the oil and also mud which can be removed by emptying the bucket.

2   The petrol interceptor will remove both petrol and oil. Both will rise to the surface of the water, the petrol will evaporate and pass out through the vent pipes.

   The oil will remain and be removed when the tanks are emptied and cleaned. The first chamber will also intercept mud which will require regular removal.

3   Glass reinforced plastic petrol interceptors may be obtained and these save time and labour in construction.

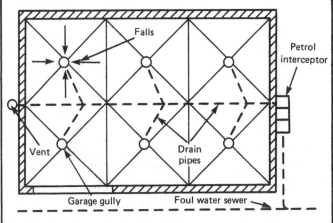

Falls

Petrol interceptor

Vent

Drain pipes

Garage gully

Foul water sewer

Fig 44 Plan of garage showing drainage

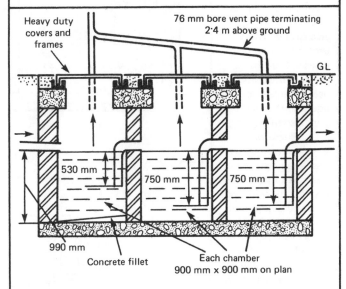

Heavy duty covers and frames

76 mm bore vent pipe terminating 2·4 m above ground

GL

530 mm

750 mm

750 mm

990 mm

Concrete fillet

Each chamber 900 mm x 900 mm on plan

Fig 45 Longitudinal section of a petrol interceptor

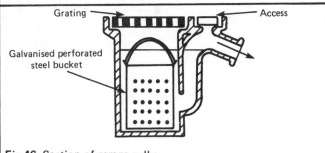

Grating

Access

Galvanised perforated steel bucket

Fig 46 Section of garage gully

DRAINAGE PUMPING — 1

Notes

1   Whenever possible drainage pipelines should be laid so that the drainage flows by gravity to the sewer or to the private sewage plant. In some cases however the site levels make it necessary to pump the drainage flows.

2   A pumping station with the motor room above or below ground level will have to be built. An electrically-driven centrifugal pump can be used and it is better if the pump is below the level of the liquid so that the pump is self priming.

3   For large schemes two pumps should be installed so that one of the pumps is standing by in case the duty pump fails. The impeller is curved on plan so that there is less risk of blockage.

4   The discharge pipe should pass into a manhole before being connected to the public sewer. This will prevent direct pumping into the sewer.

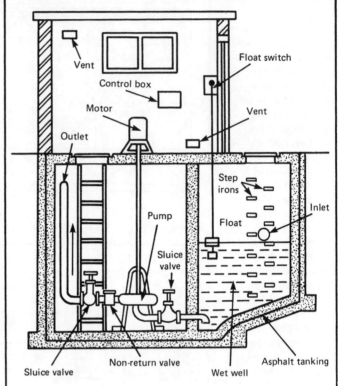

Fig 47 Section through pumping station

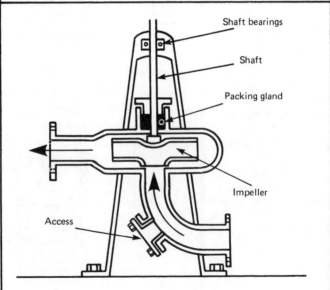

Fig 48 Section through centrifugal pump

DRAINAGE PUMPING — 2

Notes

1   The sewage ejector may be used as an alternative to the centrifugal pump for lifting foul water. The advantages of the ejector over the centrifugal pump are:
(a) Less risk of blockage.
(b) Fewer moving parts and less maintenance.
(c) A wet well is not required.
(d) One compressor unit can supply compressed air to several ejectors.

2   Operation:
(a) The incoming sewage flows through inlet pipe A into the ejector body B causing the float to rise.
(b) When the float reaches the top collar it forces the rod upwards causing an air inlet valve to open and an exhaust valve to close.
(c) Compressed air enters the ejector body and forces out the sewage through pipe C.
(d) The float falls to the bottom collar and its weight plus the rocking weight causes the air inlet valve to close and exhaust valve to open.

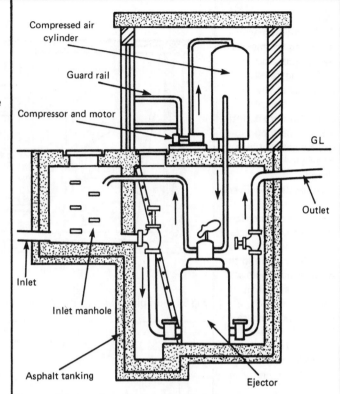

Fig 49 Section through pumping station

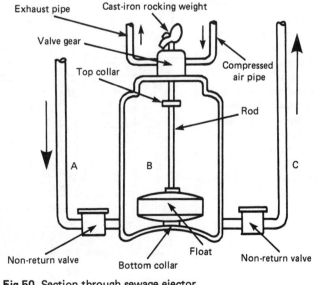

Fig 50 Section through sewage ejector

77

Notes

With any drainage pumping the makers will require the following details:

(a) Type of the drainage flow.

(b) Maximum quantity of flow per hour.

(c) Height to which the fluid has to be lifted.

(d) Length of delivery pipe.

(e) Type of electric supply (a.c. or d.c.).

1   It is sometimes necessary to build the motor room below ground level. This method is much neater and because the motor is below ground there is less noise in the vicinity of the pumping station.

2   In basements there is sometimes seepage of water which requires to be removed by a sump pump. In boiler rooms a sump pump is installed which will remove water from any leaks that may occur. It is also used for draining down the boilers.

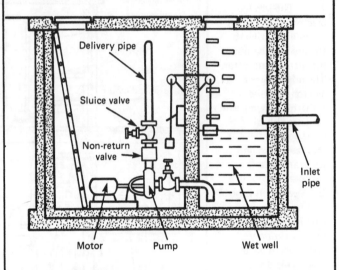

Fig 51 Pumping station with motor room below ground level

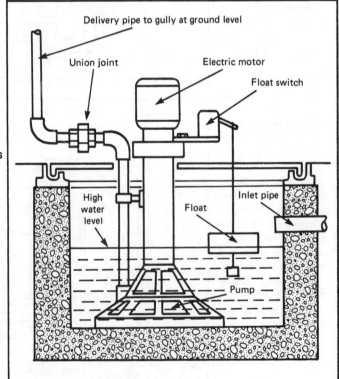

Fig 52 Sump pump

SUBSOIL DRAINAGE

Notes

1   Wherever possible a building should be constructed so that its foundations are well above the underground water table. In some cases however it is necessary to provide subsoil drainage so that the natural water table can be lowered below the building foundations.

2   A variety of systems of subsoil drainage are available and the type used will depend upon the site conditions.

3   The spacing of the branch pipes for the grid iron and herring-bone systems depends upon the nature of the soil.

4   The drain trench is 600 mm to 1·5 m deep.

5   Before the subsoil drain is discharged into a surface water drain or a water course a suitable trap must be provided.

6   Precautions must be taken to prevent back flow from the surface water drain or stream into the subsoil drain.

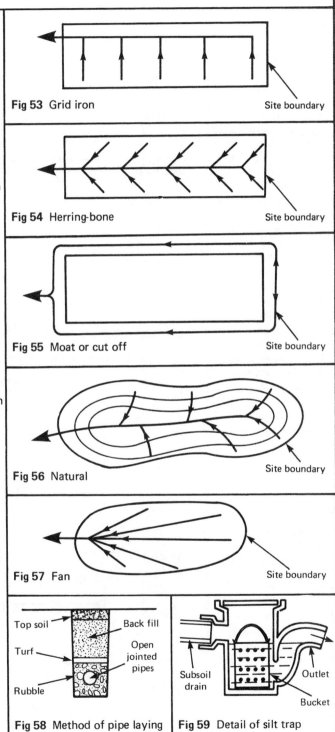

Fig 53 Grid iron   Site boundary

Fig 54 Herring-bone   Site boundary

Fig 55 Moat or cut off   Site boundary

Fig 56 Natural   Site boundary

Fig 57 Fan   Site boundary

Top soil   Back fill
Turf   Open jointed pipes
Rubble

Fig 58 Method of pipe laying

Subsoil drain   Outlet   Bucket

Fig 59 Detail of silt trap

TESTS ON DRAINS

Notes

1   The length of drain to be tested must be sealed and air pumped into the pipes until 100 mm water gauge is seen on the U gauge connected to the system. The air pressure must not fall less than 25 mm water gauge during the first 5 min.

2   The smoke test is very useful because a leak can be easily seen. The length of drain to be tested must be sealed and smoke pumped into the pipes from the lowest end. The pipes should then be inspected for any trace of smoke.

3   The water test should be carried out by inserting a stopper and then filling the pipes with water. An allowance should be made for air trapped at the joints and for water absorbed by yarn. The Building Regulations specify a water test for underground drainage. The Regulations state that a 'reasonable' water test must be applied. Minimum and maximum heads of water of 1·500 and 4·000 respectively are normally accepted.

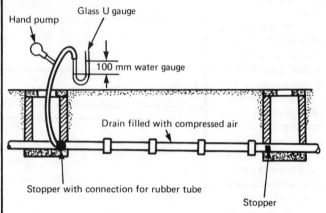

Fig 60 Air test

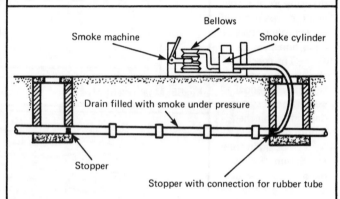

Fig 61 Smoke test

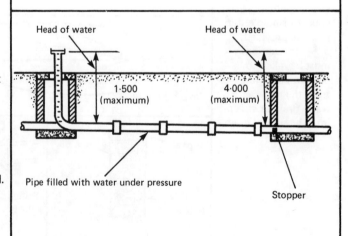

Fig 62 Water test

SOAKAWAYS

Notes

1 Where a surface water sewer is not available it will be necessary to dispose of the rainwater from the roof and other surfaces into a soakaway.

2 The soakaway must be dug in porous soil, above the water table and water from it must not flow under the building.

3 The filled soakaway is cheap to construct but its water holding capacity is greatly reduced.

4 Unfilled soakaways can be built of precast concrete brick or stone.

5 The soakaway should be sized on a basis of a rainfall intensity of 15 mm per hour. The volume of a soakaway for an area of 150 m² would be:

$$150 \times 0.015 = 2.25 \text{ m}^3$$

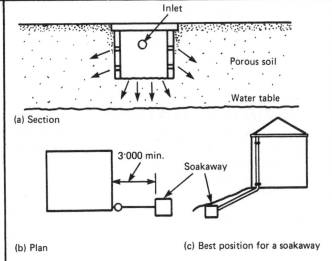

(a) Section

(b) Plan

(c) Best position for a soakaway

Fig 63 Siting of a soakaway

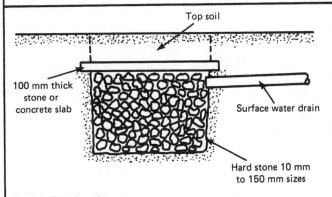

100 mm thick stone or concrete slab

Top soil

Surface water drain

Hard stone 10 mm to 150 mm sizes

Fig 64 Filled soakaway

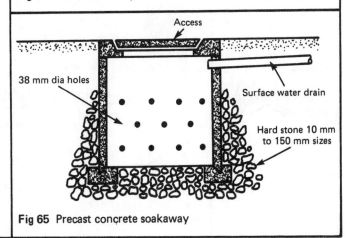

Access

38 mm dia holes

Surface water drain

Hard stone 10 mm to 150 mm sizes

Fig 65 Precast concrete soakaway

10 SOIL AND WASTE DISPOSAL SYSTEMS

LOSS OF TRAP WATER SEAL

Notes

1   Self siphonage is caused by a moving plug of water in the waste pipe causing siphonage of the trap.

2   Induced siphonage is caused by the discharge from one trap causing siphonage of another trap connected to the same waste pipe.

3   Back pressure or compression is caused when the water flowing down a stack changes direction at the bend which compresses the air in the pipe and forces out the trap water seal.

4   A piece of rag or string caught on the outlet of the trap will cause the loss of seal by capillary attraction.

5   Gusts of wind blowing across the top of a stack will cause wavering of the water and a loss of trap seal.

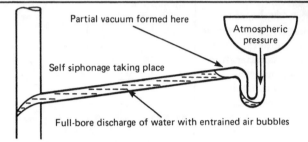

Partial vacuum formed here

Atmospheric pressure

Self siphonage taking place

Full-bore discharge of water with entrained air bubbles

Fig 1 Self siphonage

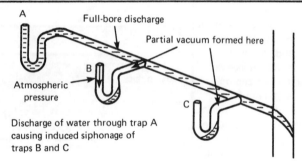

A

Full-bore discharge

Partial vacuum formed here

B

Atmospheric pressure

C

Discharge of water through trap A causing induced siphonage of traps B and C

Fig 2 Induced siphonage

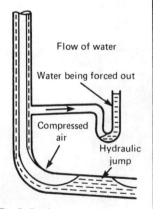

Flow of water

Water being forced out

Compressed air

Hydraulic jump

Fig 3 Back pressure or compression

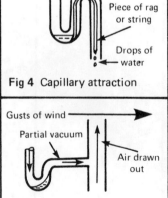

Piece of rag or string

Drops of water

Fig 4 Capillary attraction

Gusts of wind

Partial vacuum

Air drawn out

Fig 5 Wavering out

RESEALING AND ANTI-SIPHON TRAPS

Notes

Trap water seals can be maintained by fitting either a resealing or anti-siphon trap.

1   The McAlpine trap has a reserve chamber into which the water is forced when siphonage takes place. After siphonage the water in the chamber falls and reseals the trap.

2   The Grevak trap contains an anti-siphonage pipe through which air flows to break the siphonic action.

3   The Econa trap contains a cylinder on the outlet into which the water flows when the trap is siphoned. After siphonage has taken place the water in the cylinder reseals the trap.

4   The anti-siphon trap has a valve on its outlet which opens and allows air to flow into the outlet of the trap thus preventing siphonage.

5   Unfortunately resealing and anti-siphon traps require more maintenance than ordinary traps and they are liable to be noisy.

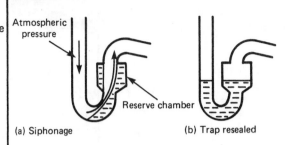

(a) Siphonage   (b) Trap resealed

Fig 6 The McAlpine resealing trap

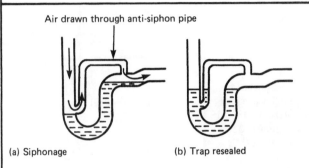

(a) Siphonage   (b) Trap resealed

Fig 7 The Grevak resealing trap

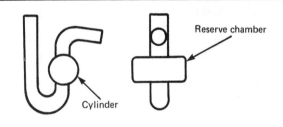

Fig 8 The Econa resealing trap

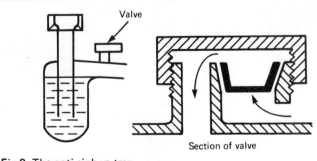

Section of valve

Fig 9 The anti-siphon trap

THE SINGLE STACK SYSTEM

Notes

1 The single stack system was introduced by the Building Research Establishment and reduces the cost of soil and waste systems. Branch vent pipes are not required but the lengths and falls of the waste pipes must be carefully governed to prevent loss of trap water seals. The trap water seals on the waste traps must be 76 mm deep.

2 The slopes of the branch pipes are: sink and bath, 18 to 19 mm/m; basin 20 to 120 mm/m; WC 18 mm/m (min).

3 The vertical stack must be straight below the highest sanitary appliance.

4 The branch bath waste must be connected to the stack 200 mm below the centre of the WC branch connection.

5 The use of an S trap type WC lowers the connection of the WC branch pipe into the stack and the 50 mm bore parallel pipe may be omitted.

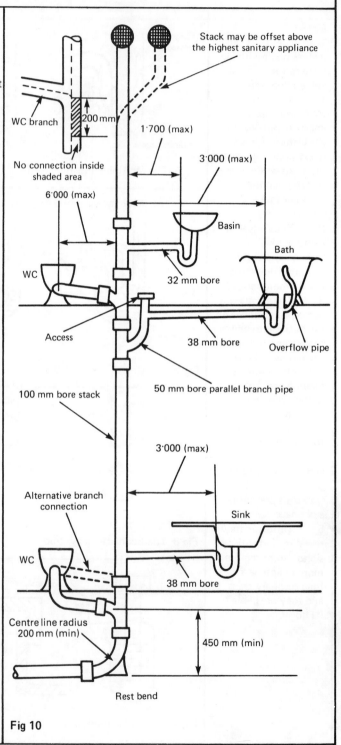

Fig 10

COLLAR BOSS SINGLE STACK SYSTEM

Notes

1   The collar boss system is a modified form of the single stack system. By use of the collar boss the restrictions that are imposed between the bath waste pipe and the stack are eliminated. The bath waste can therefore connect to the stack at a higher point without the risk of the WC discharge backing up into the bath waste pipe.

2   By installing loop vent pipes to the basin and sink traps and connecting these to the collar boss, the waste pipes from these appliances can drop vertically before running horizontally to the stack. This makes it easy to hide the waste pipes from the basin and sink.

3   The loop vent pipe on the basin trap will prevent its siphonage when the bath is discharged.

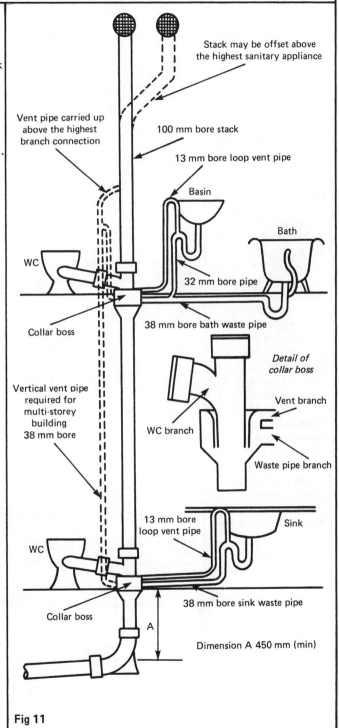

Stack may be offset above the highest sanitary appliance

Vent pipe carried up above the highest branch connection

100 mm bore stack

13 mm bore loop vent pipe

Basin

Bath

WC

32 mm bore pipe

Collar boss

38 mm bore bath waste pipe

Detail of collar boss

Vent branch

WC branch

Waste pipe branch

Vertical vent pipe required for multi-storey building 38 mm bore

13 mm bore loop vent pipe

Sink

WC

Collar boss

38 mm bore sink waste pipe

A

Dimension A 450 mm (min)

Fig 11

MODIFIED SINGLE STACK SYSTEM

Notes

1   The ventilated stack system is used in buildings where close grouping of the sanitary appliances makes it possible to install the branch waste and soil pipes without the need for individual branch ventilating pipes.

2   To prevent the loss of trap water seals the WC branch pipe must be not less than 1'00 mm bore and the angle θ between $90^1/_2{}^\circ$ and 95°.

3   To prevent the loss of trap water seals the basin main waste pipe must not be less than 50 mm bore and the angle θ between 91° and $92^1/_2{}^\circ$.

4   If the number of basins exceeds five or if the length of the main waste pipe exceeds 4·5 m a 25 mm bore vent pipe should be connected to the main waste pipe at a point between the two basins farthest from the stack.

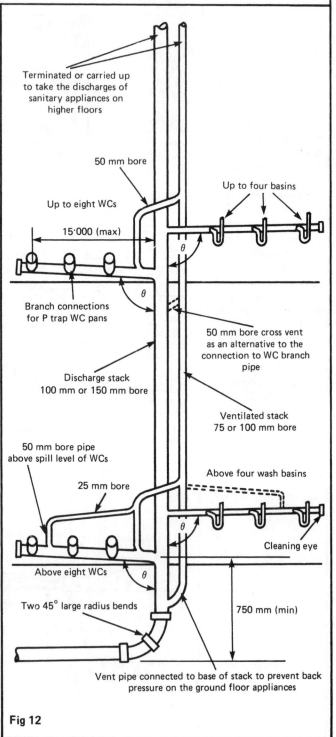

Terminated or carried up to take the discharges of sanitary appliances on higher floors

50 mm bore

Up to four basins

Up to eight WCs

15·000 (max)

Branch connections for P trap WC pans

50 mm bore cross vent as an alternative to the connection to WC branch pipe

Discharge stack 100 mm or 150 mm bore

Ventilated stack 75 or 100 mm bore

50 mm bore pipe above spill level of WCs

Above four wash basins

25 mm bore

Above eight WCs

Cleaning eye

Two 45° large radius bends

750 mm (min)

Vent pipe connected to base of stack to prevent back pressure on the ground floor appliances

Fig 12

FULLY VENTILATED ONE-PIPE SYSTEM

Notes

1   The fully ventilated one-pipe system is used in buildings where there are a large number of sanitary appliances in ranges, e.g. factories, schools, offices and hospitals.

2   Each trap is provided with an anti-siphon or vent pipe and this pipe must be connected to the discharge pipe in direction of the flow of water at a point of not nearer than 75 mm or further than 450 mm from the crown of the trap.

3   The branch vent pipe should be looped above the spill level of the sanitary appliances.

4   The vent stack should be carried down and connected to the discharge stack near to the bend. This is to remove any compressed air at this point.

5   The system is used extensively in the USA.

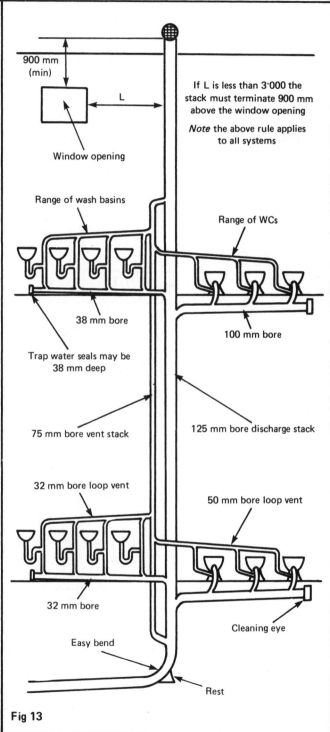

900 mm (min)

L

If L is less than 3·000 the stack must terminate 900 mm above the window opening

Note the above rule applies to all systems

Window opening

Range of wash basins

Range of WCs

38 mm bore

100 mm bore

Trap water seals may be 38 mm deep

75 mm bore vent stack

125 mm bore discharge stack

32 mm bore loop vent

50 mm bore loop vent

32 mm bore

Cleaning eye

Easy bend

Rest

Fig 13

THE TWO-PIPE SYSTEM

Notes

1   The two-pipe system is the most expensive of the soil and waste disposal systems and should therefore only be used in circumstances where the sanitary appliances are widely spaced. In buildings such as hospitals, schools, factories and even houses, wash basins or sinks may be sited in rooms some distance from the main soil stack and it is then necessary to connect these appliances to a separate waste stack.

2   The waste stack may be connected to the horizontal drain either via a rest bend or a back-inlet gully.

3   In the system, waste appliances such as basins, sinks, bidets and showers are connected to a waste stack and soil appliances such as WCs are connected to the soil stack.

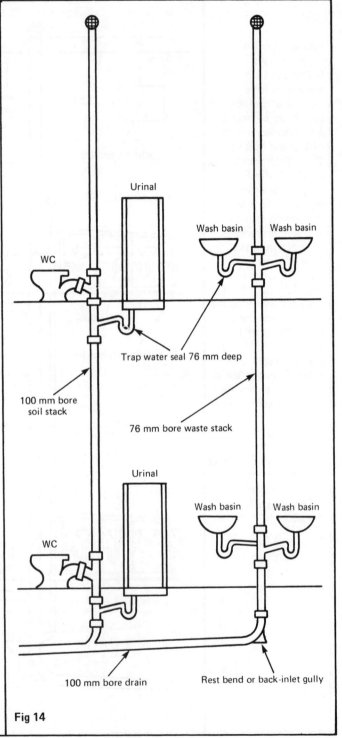

Urinal

Wash basin   Wash basin

WC

Trap water seal 76 mm deep

100 mm bore soil stack

76 mm bore waste stack

Urinal

Wash basin   Wash basin

WC

100 mm bore drain

Rest bend or back-inlet gully

Fig 14

WASTE ARRANGEMENTS FOR RANGES OF WASH BASINS

Notes

The type of waste arrangement for ranges of basins depends upon the type of building and the number of basins in the range.

1 For ranges of up to four basins branch ventilating pipes are not required providing that the bore of the main waste pipe is 50 mm and its slope is between 1° and 2½°.

2 For ranges of above four basins the bore and slope of the main pipe is the same but a 25 mm bore vent pipe is required.

3 For ranges of more than four basins resealing or anti-siphon traps may be used.

4 In schools and factories a running trap may be used providing that the length of the main waste pipe does not exceed 5 m.

5 Alternatively the wastes may discharge into a glazed channel.

6 For high class work traps may be provided with a vent or anti-siphon pipe.

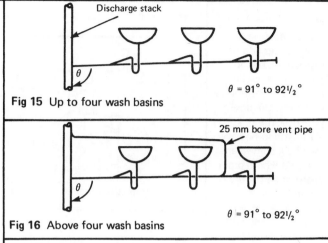

Discharge stack

$\theta = 91°$ to $92\frac{1}{2}°$

Fig 15 Up to four wash basins

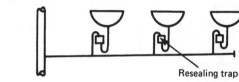

25 mm bore vent pipe

$\theta = 91°$ to $92\frac{1}{2}°$

Fig 16 Above four wash basins

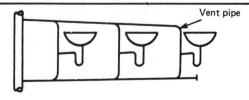

Resealing trap

Fig 17 Use of resealing or anti-siphon traps

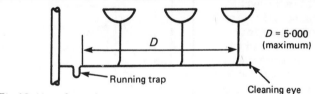

$D = 5·000$ (maximum)

Running trap

Cleaning eye

Fig 18 Use of running trap

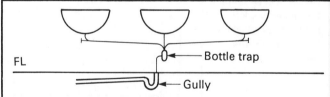

FL

Bottle trap

Gully

Fig 19 Use of bottle trap

Vent pipe

Fig 20 Use of trap ventilating pipes

WASTE PIPES FROM WASHING MACHINES AND DISH WASHERS

Notes

The cheapest and simplest method of discharging the hose-pipe from a washing machine or dishwasher is to bend the hose pipe over the sink. This method however may be inconvenient due to the hose pipe obstructing the use of the sink.

1   A tee may be inserted on the inlet side of the trap and a waste pipe then connected to this tee for the machine hose.

2   If a horizontal waste pipe is required at low level a separate waste pipe may be used and this pipe may be ventilated to atmosphere. (The vent pipe must not be connected to the ventilating stack.)

3   Alternatively, the machine hosepipe may be inserted into the vertical waste pipe and an air gap left between the machine hose pipe and the vertical waste pipe.

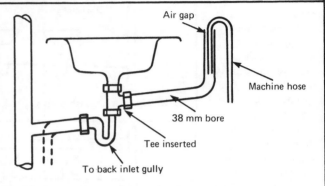

Fig 21 Connection to sink waste pipe

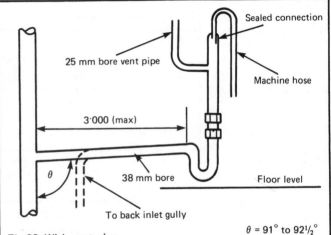

Fig 22 With vent pipe

θ = 91° to 92½°

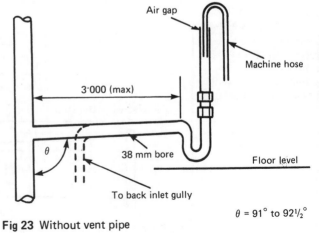

Fig 23 Without vent pipe

θ = 91° to 92½°

AIR TEST ON SOIL AND WASTE DISPOSAL SYSTEMS

Notes

1   The Building Regulations require that soil and waste disposal systems shall be capable of withstanding an air or smoke test for a minimum of three minutes at a pressure equal to a head of water of 38 mm.

2   Before testing, insert stoppers at top and bottom of the stack. Pour water over the top stopper and flush one of the WCs so that the bottom stopper is also sealed with water. Pour water in each sanitary appliance to ensure that the traps are sealed.

3   To carry out the test, pass the tube connected to the U gauge through the water seal in one of the WCs. Pump air into the pipework until the U gauge shows 38 mm water gauge. Allow a few minutes for the air temperature to stabilise. During the next three minutes the water level in the U gauge must remain stationary.

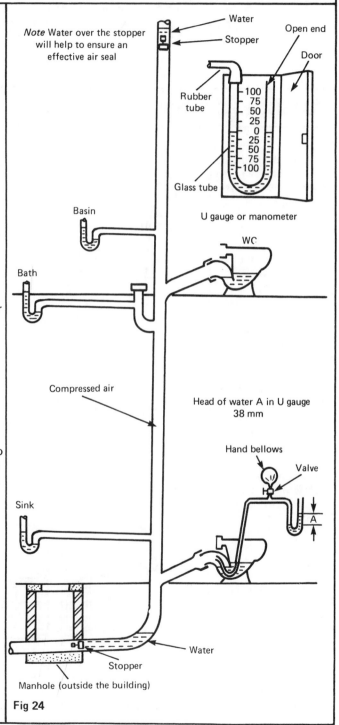

Note Water over the stopper will help to ensure an effective air seal

U gauge or manometer

Basin

Bath

Compressed air

Sink

Head of water A in U gauge 38 mm

Hand bellows

Valve

Manhole (outside the building)

Fig 24

11 FLUSHING DEVICES AND SANITARY APPLIANCES

FLUSHING CISTERNS

Notes

1   The bell-type flushing cistern is rather noisy but may be used in factories and schools. The cistern is operated by the chain being pulled down which also lifts the bell. When the chain is released the bell falls thus displacing water under the bell down the stand pipe. Siphonic action is then created which empties the cistern.

2   The disc-type flushing cistern may be used for all types of building. The cistern is operated by depressing the lever which lifts the piston and water is displaced over the siphon. Siphonic action is then created which empties the cistern.

The cistern may incorporate a dual flush siphon. When the cistern is operated in the normal manner air passing through the vent pipe breaks the siphonic action to give a $4\frac{1}{2}$ litre flush. When the lever is held down a $7\frac{1}{2}$ litre flush is obtained

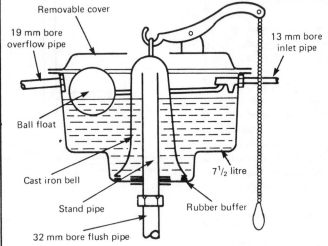

Fig 1 Bell type flushing cistern

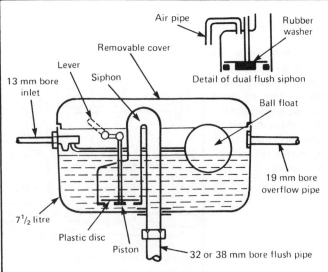

Fig 2 Disc or piston-type flushing cistern

FLUSHING TROUGH

Notes

1   A flushing trough may be used as an alternative to separate flushing cisterns for the flushing of ranges of WCs. They are used in schools factories and offices.

2   The trough saves on valves, pipework and labour, it also reduces the waiting time between consecutive flushing operations.

3   If repairs to the trough have to be carried out, it has the disadvantage of putting the range of WCs out of operation until the repair has been completed.

4   The trough may be supported on brackets or on the partition.

5   The siphon is operated in the same way as a separate cistern but as water flows through the siphon air is drawn out of the air pipe and water is thus siphoned out of the anti-siphon device. The flush is then terminated and the device refilled through the small hole.

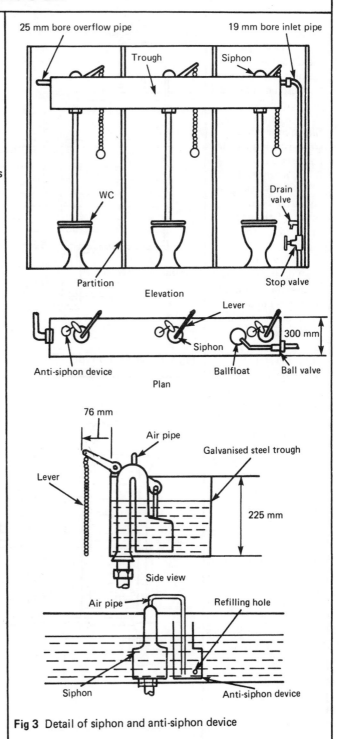

Fig 3 Detail of siphon and anti-siphon device

AUTOMATIC FLUSHING CISTERNS

Notes

1   The Roger Field's automatic flushing cistern is used for flushing of ranges of infant WCs.

2   As the cistern fills, the air in the stand pipe is gradually compressed.

3   When the head of water 'H' is slightly above the head of water 'h', the water in the trap is forced out.

4   Siphonic action is set up and the cistern flushes the WCs until air enters under the dome and breaks the siphonage.

5   With the smaller cistern, water rises inside the cistern until it reaches the air hole.

6   Air inside the dome is thus trapped and is compressed as the water rises.

7   When the water rises above the dome, the compressed air forces water out of the U tube which lowers the air pressure in the stand pipe.

8   Siphonic action is set up and the cistern is emptied.

9   The water in the reserve chamber is siphoned through the siphon tube to the lower well.

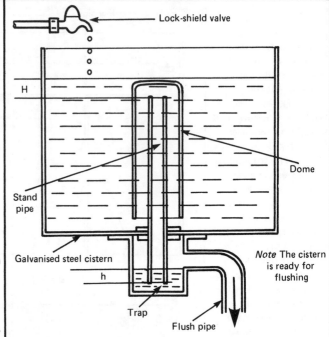

Lock-shield valve

H

Stand pipe

Galvanised steel cistern

Dome

h

Trap

Flush pipe

Note The cistern is ready for flushing

Fig 4 Roger Field's type

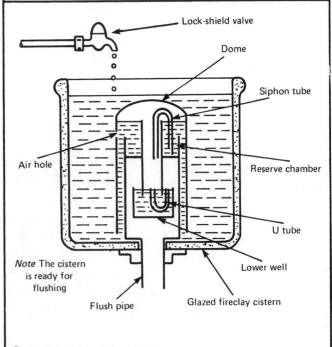

Lock-shield valve

Dome

Siphon tube

Air hole

Reserve chamber

U tube

Lower well

Flush pipe

Glazed fireclay cistern

Note The cistern is ready for flushing

Fig 5 Smaller type for urinals

AUTOMATIC FLUSHING VALVE

Notes

1   The flushing valve may be used instead of flushing cisterns or troughs. It is essential to obtain approval of the water authority before installing the valves.

2   The valves must always be supplied through a storage cistern and the minimum and maximum heads of water above the valves are 2·2 m and 36 m respectively.

3   When the flushing handle is operated the release valve is tilted and water is released from the upper chamber.

4   The greater force of water under the piston 'A' lifts valve 'B' from its seating and water flows through the outlet.

5   Water flows through the by-pass and refills the upper chamber thus cancelling out the upward force acting under piston A and valve B closes under its own weight.

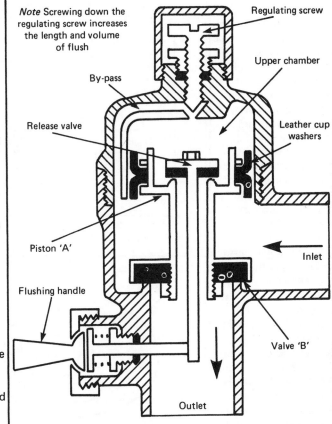

Note Screwing down the regulating screw increases the length and volume of flush

Regulating screw

Upper chamber

By-pass

Release valve

Leather cup washers

Piston 'A'

Inlet

Flushing handle

Valve 'B'

Outlet

Fig 6 Section through flushing valve

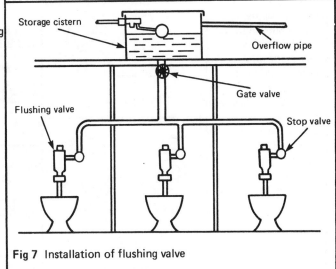

Storage cistern

Overflow pipe

Gate valve

Flushing valve

Stop valve

Fig 7 Installation of flushing valve

WASH DOWN WATER CLOSET AND JOINTS

Notes

1　The wash down WC pan is cheap simple and efficient and rarely becomes blocked. It is used in all types of buildings and is made from vitrious china, glazed fireclay or stoneware.

2　The contents of the pan are removed by the momentum of the water flush and a high-level flushing cistern gives a good flush but is noisy. A low-level cistern gives a quieter flush, is neater and now more popular.

3　The outlet of the pan may be horizontal, P, S, left or right hand.

4　A plastic connector is a popular outlet joint and only requires pushing over the outlet and into the soil-pipe collar. The flush pipe joint is usually made by a rubber cone connector.

5　Some WCs have a near horizontal outlet.

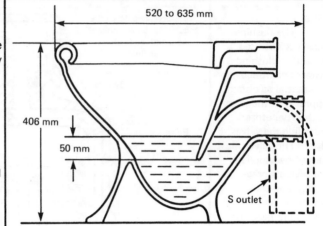

Fig 8 Section of horizontal outlet pan

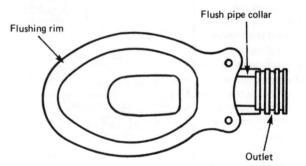

Fig 9 Plan

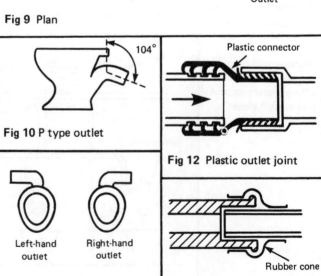

Fig 10 P type outlet

Fig 11　Left-hand outlet　Right-hand outlet

Fig 12 Plastic outlet joint

Fig 13 Rubber flush pipe joint

SIPHONIC WATER CLOSETS

Notes

1   The double trap type siphonic WCs are ideal for houses and hotel bathrooms but are not recommended for schools or factories.

2   When the cistern is flushed water flows through the pressure reducing fitting A and this reduces the air pressure in chamber B. Siphonic action is set up and the contents of the first trap are removed.

3   The first trap is refilled from the reserve chamber C.

4   The single trap type siphonic WCs should only be used in domestic buildings.

5   When the cistern is flushed the water flows through the outlet which is designed to slow down the flow of water and the pipe is thus filled with water which causes siphonic action.

6   The siphonic WCs are much quieter in operation than wash down WCs.

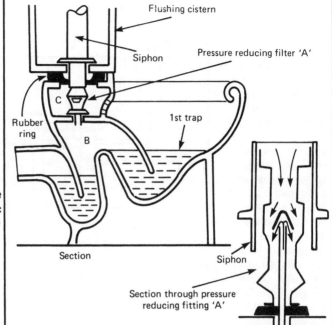

Fig 14 Double-trap type siphonic pan

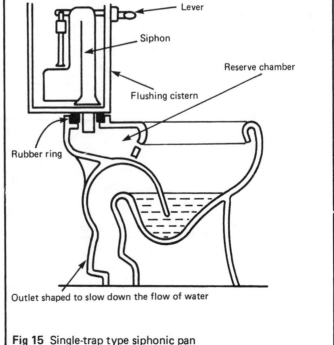

Fig 15 Single-trap type siphonic pan

BIDET

Notes

1   The bidet is classified as a waste fitting and the waste pipe therefore may be treated in the same manner as a waste pipe from a basin, sink, bath or shower.

2   The fitting is used for washing the excretory organs but it may also be used as a foot bath.

3   The hot and cold supplies are mixed at the correct temperature for the ascending spray and for greater comfort the rim of the fitting may be heated from the mixed hot and cold water supplies.

4   Because the spray nozzle is below the spill level of the fitting there is a risk of waste water being siphoned back into other draw off points. To prevent this occurrence the hot and cold water supplied to the bidet must be separated from the supplies to other fittings.

5   If the hot and cold supplies are over the rim, an air gap is required between the rim and the tap outlets.

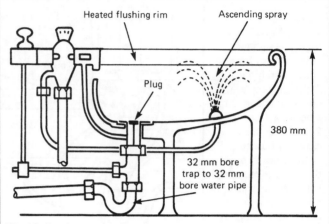

Fig 16 Section

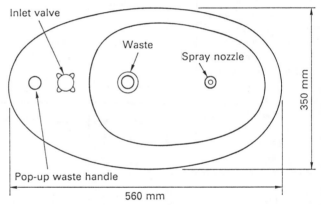

Fig 17 Plan

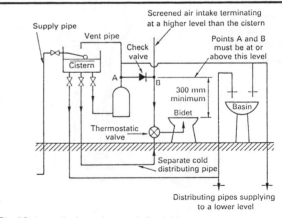

Fig 18 Installation pipework for bidet

SHOWER

Notes

1   A shower is quicker to use than a bath, is more hygienic, takes up less space and uses only about one third of the water used for a bath.

2   The mixer may be non-thermostatic or thermostatic but the latter type should be recommended to avoid risk of scalding.

3   The shower tray may be made from plastic or enamelled fireclay.

4   The bore of the supply pipes are 13 mm and they should be as short as possible to reduce frictional losses.

5   The minimum head of water above the shower outlet should be 1 m and if this is not possible a pump on the outlet pipe is required.

6   Modern thermostatic shower mixers do not require an outlet valve or non-return valves on the hot and cold supplies.

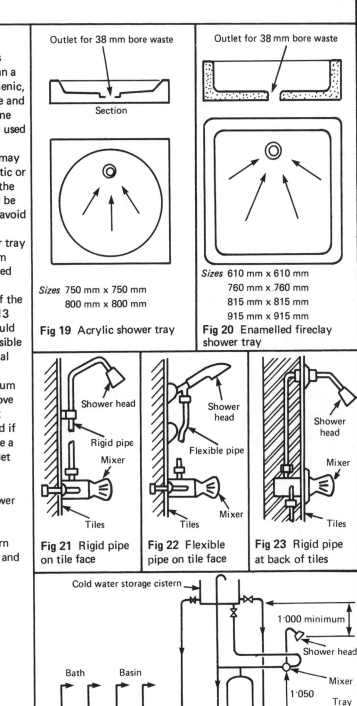

Outlet for 38 mm bore waste

Section

Sizes 750 mm x 750 mm
800 mm x 800 mm

Fig 19 Acrylic shower tray

Outlet for 38 mm bore waste

Sizes 610 mm x 610 mm
760 mm x 760 mm
815 mm x 815 mm
915 mm x 915 mm

Fig 20 Enamelled fireclay shower tray

Shower head

Rigid pipe

Mixer

Tiles

Fig 21 Rigid pipe on tile face

Shower head

Flexible pipe

Mixer

Tiles

Fig 22 Flexible pipe on tile face

Shower head

Mixer

Tiles

Fig 23 Rigid pipe at back of tiles

Cold water storage cistern

1·000 minimum

Shower head

Bath   Basin

Mixer

1·050

Tray

Fig 24 Installation pipework for shower

BATHS

Notes

1 Although a shower is a more hygienic and efficient method of washing than a bath, many people find a bath more relaxing.

2 Baths are made from acrylic sheet reinforced glass fibre, enamelled pressed steel or enamelled cast iron. The acrylic sheet bath is very popular and has the advantage of cheapness. It is also very light in weight and may be produced in various colours. The bath however must be supported on timber laid across metal cradles.

3 The corner type bath may be regarded as a luxury and the mixer taps may be placed at one corner so as to make it easier to enter or leave the bath.

4 The Sitz bath has a stepped bottom to form a seat. It may be used where space is limited or for the elderly.

Dimensions (mm)
A = 540
B = 700
C = 1700
D = 180
E = 380

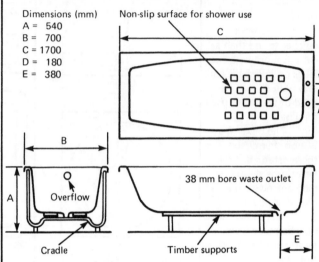

Non-slip surface for shower use

Overflow

Cradle

38 mm bore waste outlet

Timber supports

Fig 25 Acrylic sheet bath (Magna type)

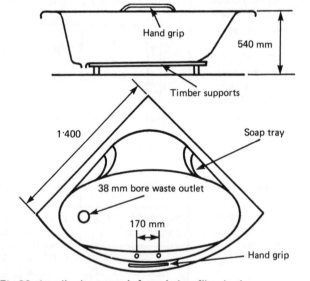

Hand grip

540 mm

Timber supports

1·400

Soap tray

38 mm bore waste outlet

170 mm

Hand grip

Fig 26 Acrylic sheet or reinforced glass fibre bath

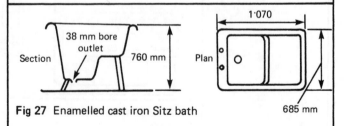

Section

38 mm bore outlet

760 mm

Plan

1·070

685 mm

Fig 27 Enamelled cast iron Sitz bath

SINKS

Notes

1   The Belfast type sink has an integral weir overflow and water may pass through this to the waste pipe via a slotted waste fitting. A draining board may be fitted at one end or two draining boards may be fitted, one at each end.

Alternatively the sink may be equipped with an integral drainer or fireclay.

2   The London sink is similar to the Belfast sink but does not have an integral overflow.

3   Sinks may also be made of enamelled pressed steel or cast iron, acrylic sheet or stainless steel.

4   The stainless steel sink may have a single bowl with left or right hand drainers or double drainers; double bowls with left or right hand drainers or double drainers.

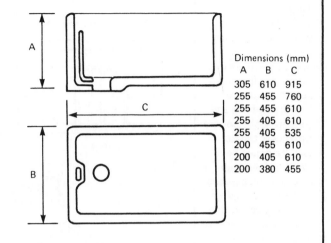

Dimensions (mm)

A	B	C
305	610	915
255	455	760
255	455	610
255	405	610
255	405	535
200	455	610
200	405	610
200	380	455

Fig 28 Enamelled fireclay Belfast sink

Dimensions (mm)

A	B	C
255	455	610
200	380	455

Fig 29 Enamelled fireclay London sink

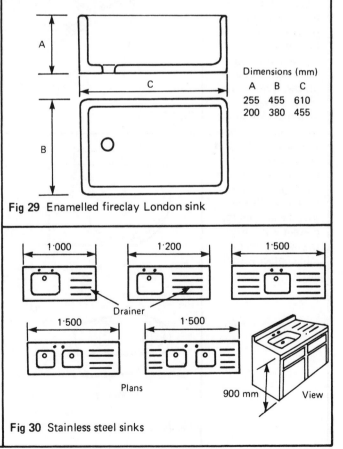

Fig 30 Stainless steel sinks

CLEANERS' SINK — DRINKING FOUNTAIN

Notes

1   The cleaners' sink is usually fitted inside the cleaners' cubicle and is used mainly for filling a bucket with water. 13 mm bore hot and cold draw-off taps are fitted over the sink and they must be fitted high enough for a bucket to be placed under them. A hinged stainless steel grating is fitted to the sink as a rest for the bucket. The grating rests on a hard wood pad fixed on the front of the sink. A 38 mm bore waste pipe is required for the sink.

2   A drinking fountain is fitted in offices, factories and schools. Several patterns are available which may be support supported on brackets or a pedestal. It is provided with a 13 mm bore non-corrosive inlet valve. The jet must be hooded to prevent water from the mouth coming into contact with the jet.

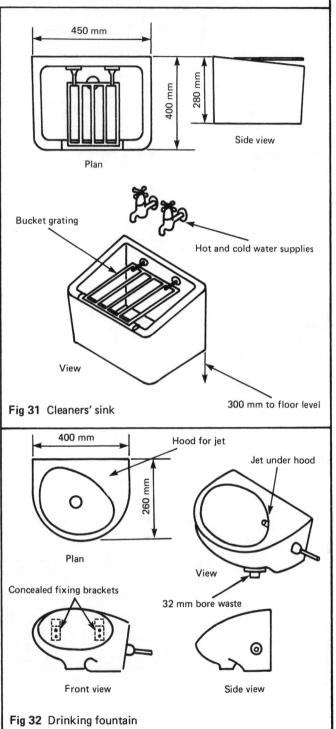

450 mm

400 mm

280 mm

Side view

Plan

Bucket grating

Hot and cold water supplies

View

300 mm to floor level

Fig 31 Cleaners' sink

400 mm

Hood for jet

Jet under hood

260 mm

Plan

View

Concealed fixing brackets

32 mm bore waste

Front view

Side view

Fig 32 Drinking fountain

102

WASH BASINS — WASHING TROUGH

Notes

1　There are a variety of types and shapes of wash basin ranging from a small hand rinse basin to a surgeon's basin. The standard basin consists of a bowl, soap tray, weir overflow and holes for taps. The basin may be supported on cast-iron brackets screwed to the wall, a corbel which is an integral part of the basin and is built into the wall, or a pedestal which conceals the pipework.

　Basins are made from vitreous china, glazed fireclay plastic or enamelled cast iron.

2　A washing trough may be installed as an alternative to a range of wash basins. The trough may have draw-off taps or an umbrella spray. The water supply to the taps or spray is thermostatically controlled to provide the correct temperature for hand washing of 45°C.

3　The trough saves on space and takes less time to install than a range of basins. It is also very hygienic.

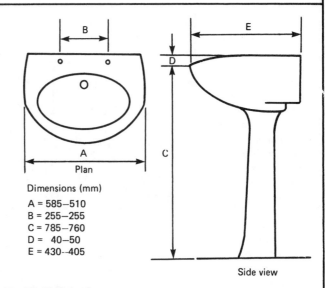

Dimensions (mm)
A = 585—510
B = 255—255
C = 785—760
D = 40—50
E = 430—405

Fig 33 Wash basin

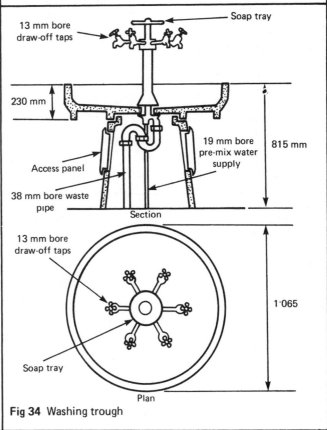

Fig 34 Washing trough

URINALS

Notes

1   Urinals are installed in commercial, industrial buildings and schools, and reduce the number of WCs in male sanitary rooms. Three types are available:

(i) The bowl type, which is screwed to the wall and if two or more bowls are fitted division pieces may be fixed between them.

(ii) The slab type which consists of flat slabs fixed against the wall, projecting return end slabs and channel.

(iii) The stall type which consists of curved stalls, dividing pieces and glazed channel.

2   Urinals are flushed at intervals of 20 minutes by means of an automatic flushing cistern at the rate of 4·5 litres per bowl, 610 mm of slab width or stall. The water to the cistern should be automatically shut off during night time and weekends.

3   A hydraulically operated inlet valve to the automatic flushing cistern is obtainable which closes when the building is unoccupied and other fittings not used.

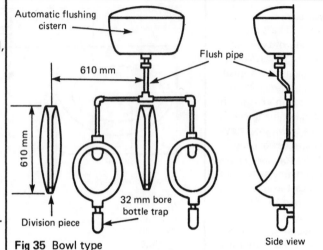

Fig 35 Bowl type

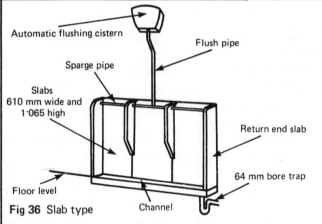

Fig 36 Slab type

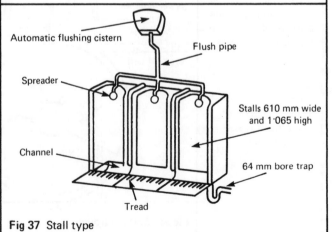

Fig 37 Stall type

HOSPITAL SANITARY APPLIANCES

Notes

1   Special types of sanitary appliances are required for hospital sluice rooms. The slop hopper is required for the efficient disposal of slops and is similar in design to the washdown WC pan but is provided with a hinged stainless steel grating for a bucket rest. Another grating inside the pan prevents the entry of large objects which could cause a blockage.

2   The bed pan washer is provided with a nozzle fitted inside the pan which produces a jet of water for the cleansing of bed pans and urine bottles. To prevent the risk of contamination of the water supply it is essential that the water for the jet is taken from the cold water storage cistern through a separate pipe. An 89 mm bore outlet is provided for the pan.

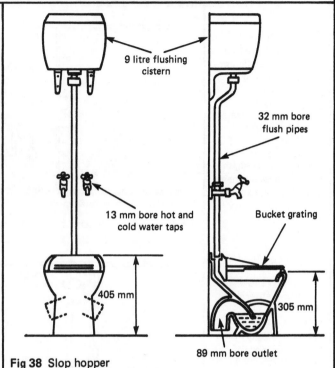

9 litre flushing cistern

32 mm bore flush pipes

13 mm bore hot and cold water taps

Bucket grating

405 mm

305 mm

89 mm bore outlet

Fig 38 Slop hopper

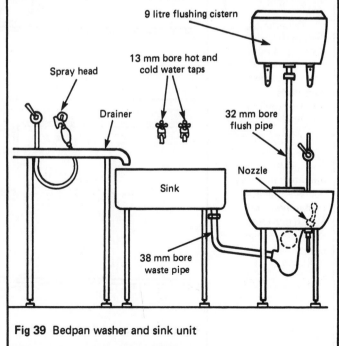

9 litre flushing cistern

13 mm bore hot and cold water taps

Spray head

Drainer

32 mm bore flush pipe

Nozzle

Sink

38 mm bore waste pipe

Fig 39 Bedpan washer and sink unit

SANITARY CONVENIENCES

Notes

1   The Building Regulations 1991 require that sufficient sanitary conveniences shall be provided which shall be:
(a) in rooms separated from places where food is stored or prepared; and (b) designed and installed so as to allow effective cleaning. Fig 40 shows the method of entry into a sanitary convenience from a kitchen via an intervening ventilated space.

2   Fig 41 shows the method of entering a bathroom via a corridor. If required the bathroom may be entered directly from a bedroom, especially if there is another WC in the dwelling. See Fig 42.

3   The Regulations require that any dwelling should have at least one WC and one wash basin. The wash basin should be located in the room containing the WC or in a room or space giving direct access to the room containing the WC (provided that it is not used for the preparation of food).

4   If mechanical ventilation is used it is required to run, at the rate of three air changes per hour, for 15 minutes after the room has been vacated.

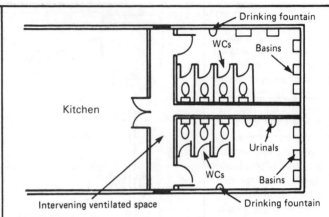

Fig 40 Sanitary accommodation from a kitchen

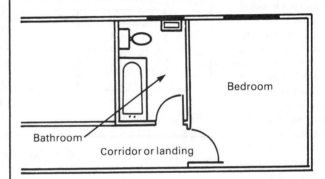

Fig 41 Entry to a bathroom via a corridor or landing

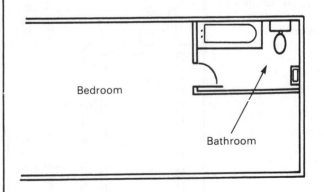

Fig 42 Entry to a bathroom directly from a bedroom

12 SEWAGE DISPOSAL

BRICK OR CONCRETE SEPTIC TANK

Notes
1   When sewage disposal is not provided by the local authority it is necessary to install a private sewage disposal plant. This is often necessary for buildings in rural areas.

2   A septic tank system is a popular method of providing a sewage disposal system. The tank is a watertight chamber in which the sewage is liquified by the action of anaerobic bacteria. This type of bacteria live in the absence of oxygen and the tank must therefore be covered.

3   The tank is divided into two chambers and its total length should be three times its breadth.

4   The sewage is finally purified by passing it either through a biological filter or subsoil drainage pipes.

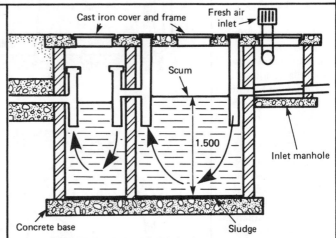

Fig 1 Longitudinal section of septic tank minimum volume under Building Regulations = 2.7 m³

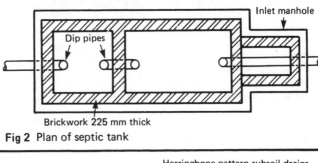

Fig 2 Plan of septic tank

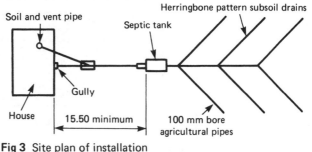

Fig 3 Site plan of installation

TREATMENT OF SEPTIC TANK EFFLUENT

Notes

1   In the biological filter the liquid from the septic tank comes into contact with a suitable medium such as hard broken stone, clinker, coke or polythene shingle.

2   The surfaces of the medium become coated with an organic film which assimilates and oxidises the polluting matter through the agency of aerobic bacteria.

3   This type of bacteria live in the presence of oxygen. The filter therefore requires an efficient system of ventilation by means of under-drains leading to vertical vent pipes.

4   Alternatively the liquid from the septic tank may be passed through a system of subsoil drains. The subsoil should be porous and at a level above that of the water table in winter.

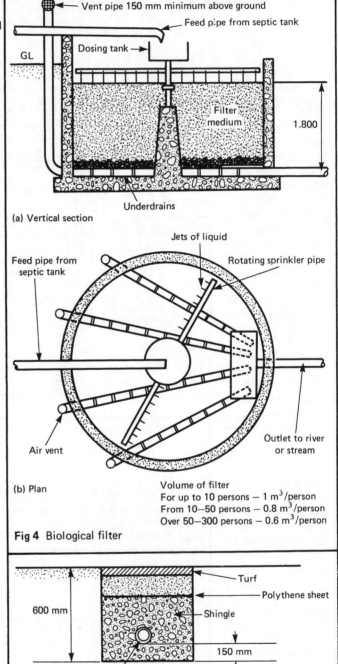

Vent pipe 150 mm minimum above ground

Feed pipe from septic tank

Dosing tank

GL

Filter medium

1.800

Underdrains

(a) Vertical section

Jets of liquid

Feed pipe from septic tank

Rotating sprinkler pipe

Air vent

Outlet to river or stream

(b) Plan

Volume of filter
For up to 10 persons — 1 m³/person
From 10–50 persons — 0.8 m³/person
Over 50–300 persons — 0.6 m³/person

Fig 4 Biological filter

Turf

Polythene sheet

600 mm

Shingle

150 mm

Open-jointed drain pipes

Fig 5 Subsoil irrigation pipe trench

KLARGESTER SETTLEMENT TANK

Notes

1   The Klargester settlement tank in glass reinforced plastics is a simple, reliable and cost effective septic tank sewage disposal system.

2   The tank may be used for a single house, a group of houses, factories or restaurants and the capacities range from 2700 litres to 100 000 litres.

3   The sewage flows through the tank through three separate chambers where it is liquefied by the action of anaerobic bacteria.

4   Sludge forms in the bottom chamber and this must be removed at least every twelve months.

5   The liquid which flows from the tank requires treatment by passing it either through subsoil irrigation pipes or a biological filter.

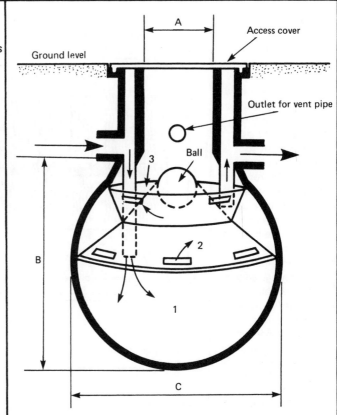

Fig 6 Section through tank

Details of Klargester tank

Capacity of tank in litres	Number of users with flow rate per head per day		Nominal dimensions in mm.		
	180 litres	250 litres	A	B	C
2700	4	3	610	1850	1800
3750	9	7	610	2060	2000
4500	14	10	610	2150	2100
6000	22	16	610	2400	2300
7500	30	22	610	2630	2500
10000	44	32	610	2800	2740

Note: The floating ball will push away to give access into the lowest chamber for sludge removal

Fig 7 Dimensions of tank

THE BIODISC SEWAGE TREATMENT PLANT

Notes

1 Crude sewage enters the biozone chamber via a deflector box which slows down the flow.

2 The heavier solids sink to the bottom of the compartment to disperse into the main sludge zone and the lighter solids still in suspension pass to the biozone chamber.

3 In the biozone chamber the microorganisms present in the sewage adhere to the partially immersed rotating discs to form a biologically active film feeding upon the impurities and rendering them inoffensive.

4 Baffles separate the series of rotating, fluted discs so that the sewage flows through each disc in turn.

5 The sludge from the primary settlement zone must be removed every six months.

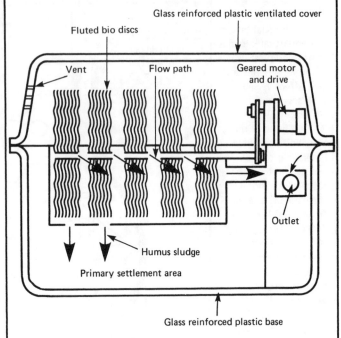

Fig 8 Longitudinal section

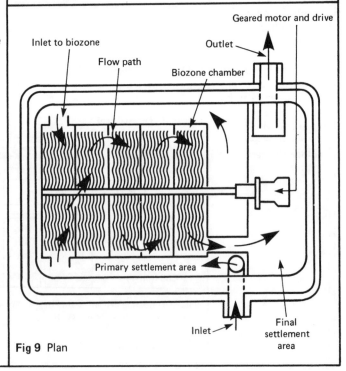

Fig 9 Plan

110

CESSPOOLS

1   A cesspool is a watertight chamber sited below ground level, for the reception and storage of sewage until such time that it is pumped out and taken away for disposal.

2   A cesspool may be constructed of brickwork rendered inside with waterproof cement mortar, pre-cast concrete rings supported on a concrete base, or glass reinforced polyester (GRP).

3   The Building Regulations require a cesspool to have a minimum volume, measured below the level of the inlet, of 18 m³.

4   A cesspool must be impervious to rainwater, well ventilated and have no outlet or overflow pipe. It should be sited at least 15.50 from a dwelling house and, if possible, built on a sloping site below the dwelling.

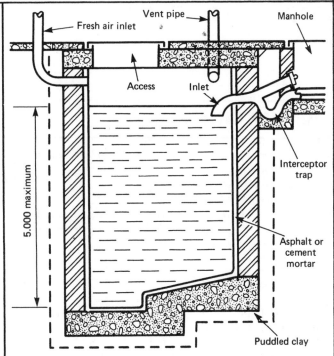

Fig 10 Brick cesspool

Capacities and lengths
18180 litres   4600 mm
27280   "   6450 mm
36370   "   8300 mm

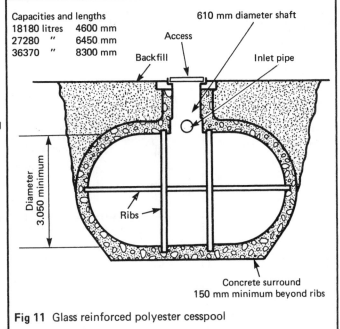

Fig 11 Glass reinforced polyester cesspool

13 REFUSE DISPOSAL

REFUSE CHUTE

Notes

1 The number and siting of refuse chutes depends upon: (a) the layout of the building; (b) the type of storage and collection; (c) volume of refuse; (d) access for refuse vehicle.

2 The chute should be sited away from habitable rooms and not more than 30 m, measured horizontally, from each dwelling.

3 It is cheaper to provide space for additional storage beneath the chute than to provide additional chutes.

4 The internal surface of the chute must be smooth and impervious to moisture. It should be constructed of refactory material that will provide at least one hour fire resistance.

5 The refuse chamber should also be of refractory material that will provide at least one hour fire resistance.

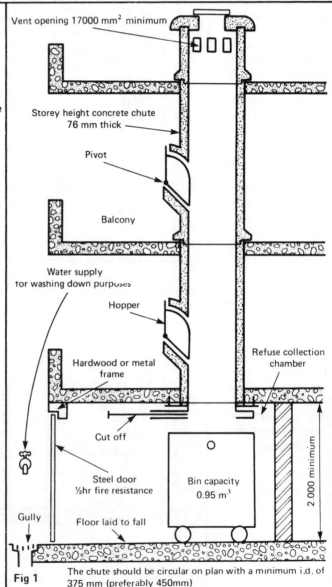

Vent opening 17000 mm² minimum

Storey height concrete chute 76 mm thick

Pivot

Balcony

Water supply tor washing down purposes

Hopper

Refuse collection chamber

Hardwood or metal frame

Cut off

Steel door ½hr fire resistance

Bin capacity 0.95 m³

Gully

Floor laid to fall

2.000 minimum

Fig 1 The chute should be circular on plan with a minimum i.d. of 375 mm (preferably 450mm)

ON-SITE INCINERATION OF REFUSE

Notes

1   In the system the flue takes the products of combustion to the roof level and an electrically-driven fan ensures suction in the refuse chute. Smoke and fumes therefore are prevented from entering the chute.

2   A large combustion chamber receives and stores the refuse until it is ignited by an automatic burner, and burning periods are thermostatically and time controlled.

3   The waste gases are washed and cleaned before being discharged into the flue. There is no restriction on wet or dry materials and glass, metal or plastics may be charged.

4   Health risk of putrifying rubbish is entirely eliminated because the residue is odourless and sterile.

5   The cost and time of refuse removal are reduced because the residue is only 7% to 12% of the initial volume.

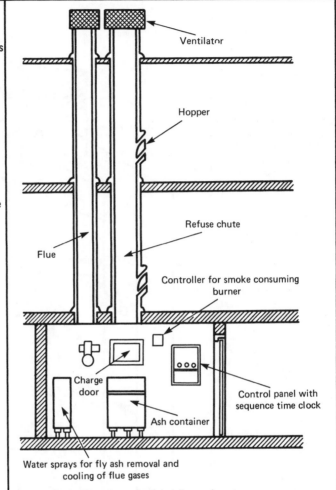

Ventilator

Hopper

Refuse chute

Flue

Controller for smoke consuming burner

Charge door

Control panel with sequence time clock

Ash container

Water sprays for fly ash removal and cooling of flue gases

Fig 2 Vertical section of refuse disposal system

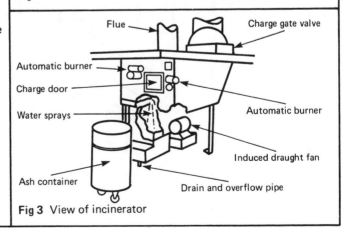

Flue

Charge gate valve

Automatic burner

Charge door

Water sprays

Automatic burner

Ash container

Induced draught fan

Drain and overflow pipe

Fig 3 View of incinerator

SANITARY INCINERATION

Notes

1   Incinerators are the quickest, easiest and most hygienic method of disposal of dressings, swabs and sanitary towels. They are usually installed in offices, hospitals and hotels.

2   When the incinerator door is opened, gas burners automatically ignite and burn the contents. After a predetermined time the gas supply is cut-off by a time switch and each time the door is opened the time switch reverts to its original position to give another burning cycle.

3   The incinerators are provided with a removable ash pan and a flame-failure safety device.

4   A fan draws the products of combustion from the incinerator and an automatic device ensures that burning does not take place without fan operation.

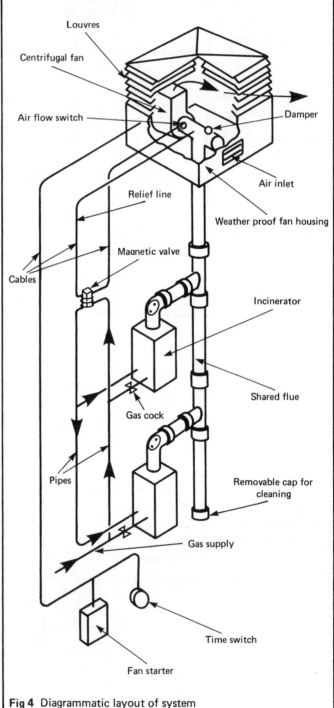

Louvres

Centrifugal fan

Air flow switch

Damper

Air inlet

Relief line

Weather proof fan housing

Magnetic valve

Cables

Incinerator

Shared flue

Gas cock

Pipes

Removable cap for cleaning

Gas supply

Time switch

Fan starter

Fig 4 Diagrammatic layout of system

THE MATTHEW-HALL GARCHEY SYSTEM

Notes

1   In the Matthew-Hall Garchey sytem of refuse disposal food waste, bottles, cans and cartons are disposed of at the source without the need to grind or crush the refuse.

2   A bowl beneath the sink retains the normal waste water. Refuse is placed inside a central tube in the sink and when the tube is raised the waste water and the refuse are carried away down a stack pipe to a chamber at the base of the building.

3   The refuse from the chamber is collected at weekly intervals by a special tanker in which the refuse is compacted into a damp, semi-solid mass that is easy to tip.

4   One tanker can collect the refuse from up to 200 dwellings in one load.

5   The waste water from the tanker is forced into the foul water sewer.

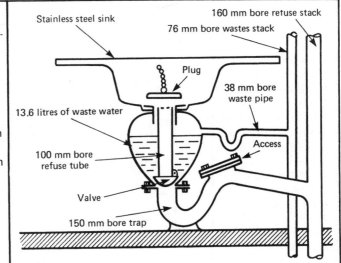

Fig 5 Detail of special sink unit

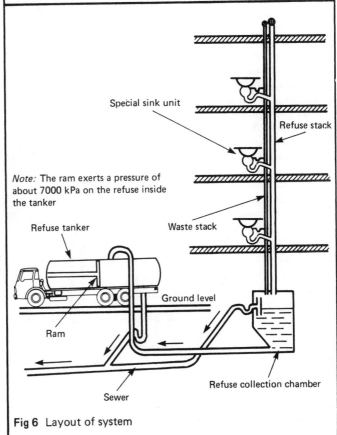

Note: The ram exerts a pressure of about 7000 kPa on the refuse inside the tanker

Fig 6 Layout of system

115

PNEUMATIC TRANSPORT OF REFUSE

Notes

1   In the pneumatic transport of pulverised refuse developed by the Building Research Establishment the refuse is disintegrated in a grinder to pieces approximately 10 mm across.

2   The refuse is then blown a short distance down a 75 mm bore pipe and lies in the pipe until, at predetermined times, a valve leading to a 150-300 mm bore pipe opens and the small pieces of refuse are picked up by an air stream.

3   The refuse is then carried by the 150-300 mm bore pipe about 1 km or more to a collection point where it is transferred to a positive pressure pneumatic system and blown to a treatment plant.

4   The system can be adapted to segregate salvagable materials, such as metal, glass and paper.

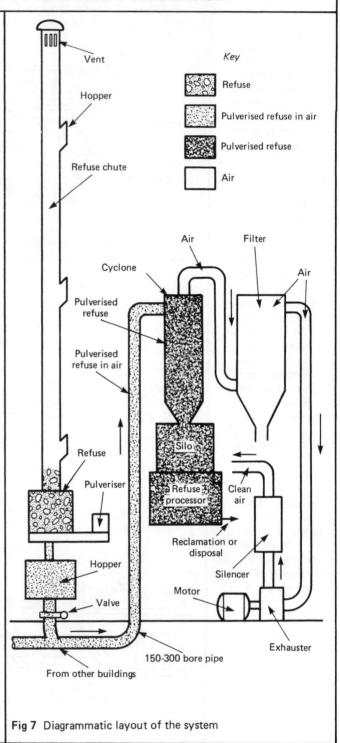

Key

Refuse

Pulverised refuse in air

Pulverised refuse

Air

Fig 7 Diagrammatic layout of the system

FOOD WASTE DISPOSAL UNITS

Notes

1   Food waste disposal units are designed to dispose of organic food waste in dwelling houses and canteen kitchens. The units must not be used to dispose of glass, metal, rags or plastics.

2   Where the chute or the Garchey system are not installed the units improve the service given by dustbins or refuse bags.

3   Food waste is fed through the sink waste to the unit in which a grinder, powered by a small electric motor, cuts the food into fine particles that are then washed down with waste water from the sink.

4   The partially liquified food particles discharge through a 38 mm (min:) bore waste pipe to a back-inlet gulley.

5   It is essential that the unit is electrically earthed.

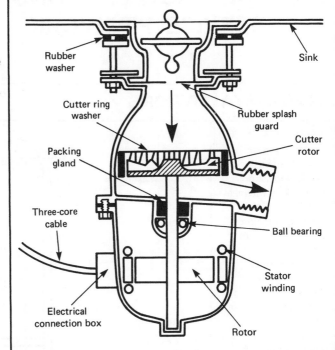

Rubber washer

Sink

Cutter ring washer

Rubber splash guard

Cutter rotor

Packing gland

Three-core cable

Ball bearing

Stator winding

Electrical connection box

Rotor

Fig 8 Section through unit

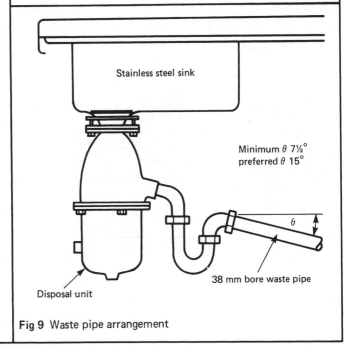

Stainless steel sink

Minimum θ 7½°
preferred θ 15°

θ

Disposal unit

38 mm bore waste pipe

Fig 9 Waste pipe arrangement

14 GAS INSTALLATIONS AND SERVICE PIPEWORK

GAS INSTALLATIONS

Notes

1   Wherever possible the gas service pipe should fall back to the main so that any water entering the pipe will drain back to the main.

2   For small installations a service pipe having a minimum diameter of 25 mm is satisfactory. A stop valve on the underground service pipe is normally only required for large buildings.

3   The underground service pipe may be of mild steel which must be protected from corrosion by wrapping with 'Denso' tape or coating with bitumen or PVC.

4   If the service pipe below ground contains a trap, the bottom of the trap must be provided with a condensate receiver and any water entering the receiver must be pumped out.

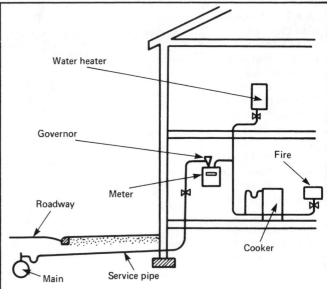

Fig 1 Typical house installation

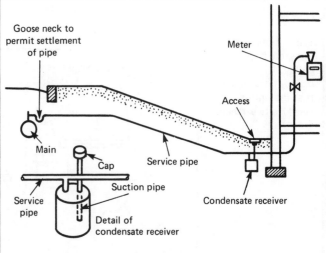

Fig 2 Use of condensate receiver

GAS SERVICE PIPE INTO BUILDINGS — 1

Notes

1   A service pipe is a pipe between the gas main and the primary meter control. Wherever possible, the service pipe should enter the building on the gas main side. The pipe must not pass under the foundation or under the base of walls or footings of a building.

2   A service pipe must not be run within a cavity or pass through it except by the shortest possible route.

3   A service pipe must not be installed in an unventilated void space but ventilated suspended floors may be accepted.

4   Where a service pipe passes through a wall or solid floor it must be enclosed by a sleeve to protect it from differential movement.

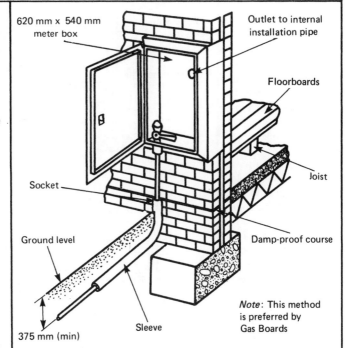

620 mm x 540 mm meter box

Outlet to internal installation pipe

Floorboards

Joist

Socket

Ground level

Damp-proof course

375 mm (min)

Sleeve

Note: This method is preferred by Gas Boards

Fig 3 Entry to an external meter box

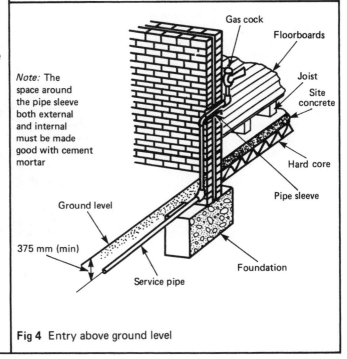

Gas cock

Floorboards

Joist

Site concrete

Note: The space around the pipe sleeve both external and internal must be made good with cement mortar

Hard core

Pipe sleeve

Ground level

375 mm (min)

Service pipe

Foundation

Fig 4 Entry above ground level

GAS SERVICE PIPE INTO BUILDINGS — 2

Notes

1   Where construction difficulties preclude the use of an external meter box or external riser, the service pipe may pass under either a solid or hollow floor.

2   For a solid floor, a sleeve should be provided built into the wall and extending into a 300 mm x 300 mm pit in the floor through which the service pipe will be laid to the meter position. The duct must have a maximum length of 2 m and must be continuous throughout.

3   The annulus between the duct and the service pipe must be sealed and the 300 mm x 300 mm hole in the floor filled with sand. The floor surface should be made good to match the floor finish.

4   Service pipes must only be laid under ventilated suspended floors and the pipe must not be more than 2 m long.

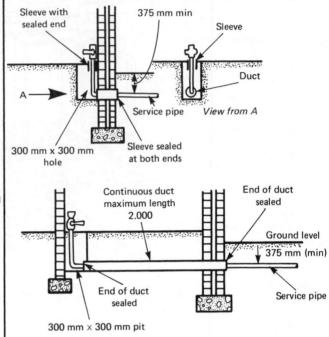

Fig 5 Service pipe entry into solid floor

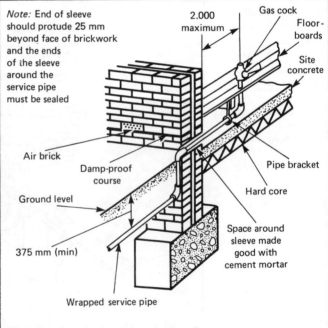

Fig 6 Service pipe entry into hollow floor

GAS SERVICE PIPE IN MULTI-STOREY BUILDINGS

Notes

1   Gas service pipe risers must be installed in protected shafts constructed in accordance with the Building Regulations.

2   Alternative methods of building a protected shaft are:

(a) A continuous shaft ventilated to the outside at top and bottom. In this case a sleeve is required where a horizontal pipe passes through the wall of the shaft;

(b) A shaft which is fire stopped at each floor level. Ventilation to the outside is required at both high and low level from each isolated section.

(c) The shaft must have a minimum fire resistance of one hour and the access door or panel a minimum fire resistance of half an hour.

(d) The gas riser must be well supported at its base.

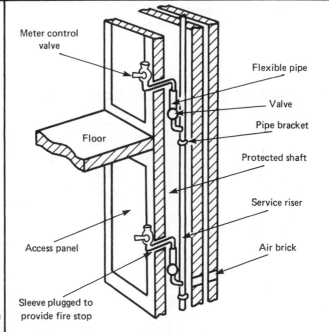

Fig 7 Service pipe in a continuous shaft

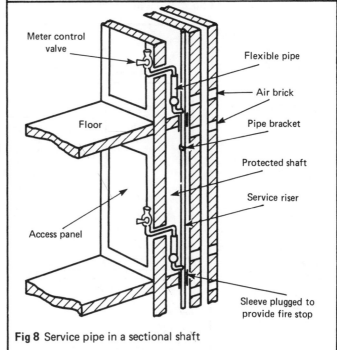

Fig 8 Service pipe in a sectional shaft

15 GAS METERS AND CONTROLS

INSTALLATION OF GAS METERS

Notes
1   The gas meter and its associated controls is the property of British Gas plc. It should be sited as close as possible to where the service pipe enters the building.
2   The meter should be accessible for maintenance, inspection and reading. It may be installed in a cupboard designed for the purpose.
3   The meter area must be well ventilated and the meter must be protected from damage, corrosion and heat.
4   A constant pressure governor must be fitted to the pipework, usually on the inlet.
5   For large industrial meters a by-pass pipe is installed with a sealed by-pass valve. The by-pass pipe may be used during repair or maintenance of the meter.

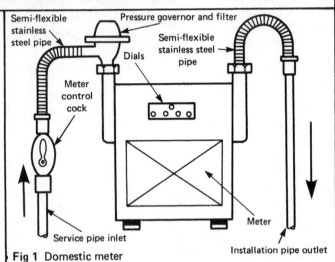

Fig 1 Domestic meter

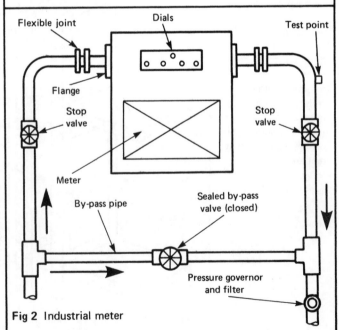

Fig 2 Industrial meter

GAS CONTROLS

Notes

1   A constant pressure governor is fitted to give the correct pressure of gas, either in the pipework or at the appliance burner.

2   Gas passes through the valve and also through the by-pass to the space between two diaphragms.

3   The main diaphragm is loaded by a spring and the upward and downward forces acting upon this diaphragm are balanced. The compensating diaphgram stabilises the valve.

4   Any fluctuation of inlet pressure inflates or deflates the main diaphragm and raises or lowers the valve thus giving a constant outlet pressure.

5   A meter control cock has a tapered plug which fits into a tapered body.

6   The drop-fan safety cock prevents a gas cock being turned on accidentally.

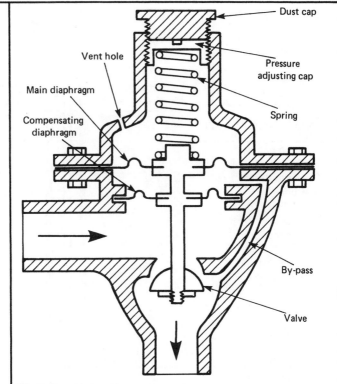

Fig 3 Constant pressure governor

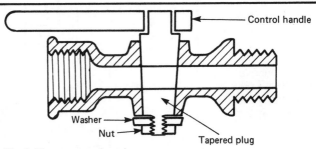

Fig 4 Meter control cock

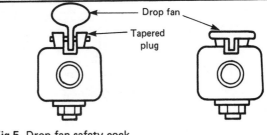

Fig 5 Drop-fan safety cock

123

GAS THERMOSTATS

Notes

1   A thermostat is a device which closes or opens a gas valve in accordance to the temperature it senses.

2   The rod-type thermostat operates on the difference in expansion between brass and invar steel when heated. When water or air surrounding a brass tube becomes hot the tube expands and carries with it an invar steel rod and closes a gas valve. The brass expands more than the invar steel and this permits a regulated movement of the rod.

3   In the vapour expansion thermostat, a bellows, capillary tube and probe are filled with ether. When the water or air surrounding the probe becomes hot, the vapour expands and in turn the bellows expands and closes a valve.

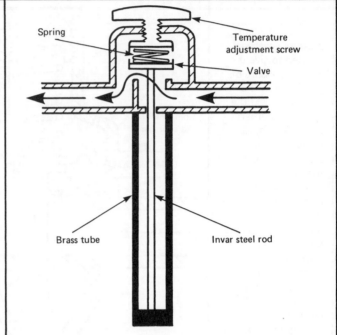

Fig 6 Rod-type thermostat

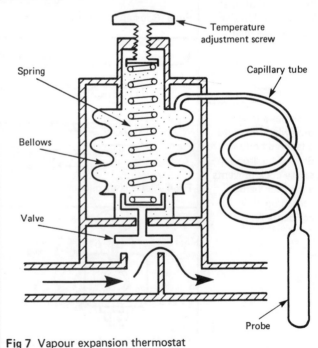

Fig 7 Vapour expansion thermostat

GAS BOILER THERMOSTAT AND RELAY VALVE. HEATER CONTROLS

Notes

1 A rod-type thermostat is often connected to a relay valve to give on/off control of the gas supply to the burner.

2 When the boiler or air heater is in operation, gas flows to the burner because valves A and B are open and the gas pressures above and below the diaphragm are equal.

3 When the water or air reaches the required temperature the brass casing of the rod thermostat expands and pulls the invar rod with it thus closing the valve A. This prevents gas from flowing to the underside of the diaphragm.

4 The gas pressure above the diaphragm builds up and allows the valve B to fall under its own weight valve B is therefore closed.

5 Various other controls are required for boilers and air heaters. The arrangement of controls are shown in Fig. 9.

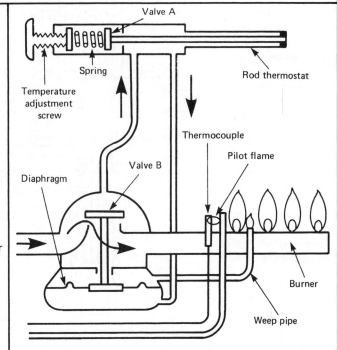

Fig 8 On/off thermostat and relay valve

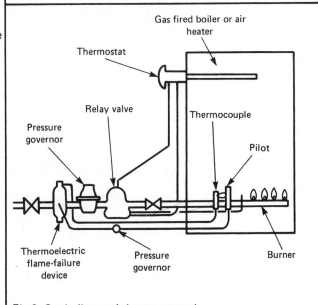

Fig 9 Gas boiler or air heater controls

FLAME FAILURE SAFETY DEVICES

Notes

1   In the thermo-electric flame failure safety device the sensing element is a thermocouple which consists of two dissimilar metals joined together at each end to form an electrical circuit. When the thermocouple is heated, a small electric current is generated and this energises an electo-magnet which holds a cut-out valve open.

2   If the pilot flame is extinguished the thermocouple cools and the electric current is no longer produced to energise the magnet and the cut-out valve closes.

3   In the bi-metal strip flame-failure device, brass and invar steel, which have different expansion rates, are joined together.

4   If the pilot flame is extinguished the bi-metal strip opens to its original position, before being heated, and this closes the inlet valve.

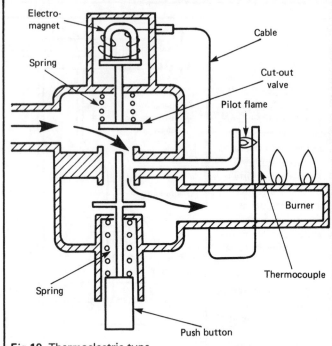

Fig 10 Thermoelectric type

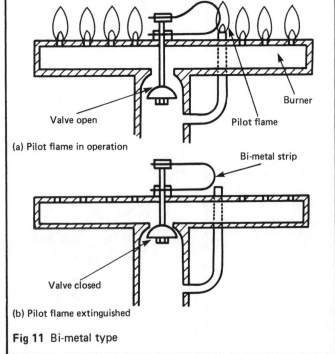

(a) Pilot flame in operation

(b) Pilot flame extinguished

Fig 11 Bi-metal type

16 GAS BURNERS AND FLUES

GAS BURNERS

Notes

1   Natural gas has a very slow burning velocity and there is a tendency for a flame to lift off the burner. A burner must be designed so that the velocity of the gas-air mixture issuing from the port is at about the same velocity of the flame.

2   If the gas pressure is too low, or the injector bore too large insufficient air is drawn into the burner and the flame is smoky and floppy.

3   If the gas pressure and injector bore are correct, sufficient air is drawn into the burner but the velocity of the gas-air mixture is too high and the flame lifts off the burner. A retention flame will prevent this flame lift off.

4   Flame lift-off may also be prevented by increasing the number of parts which slows down the velocity of the gas-air mixture. A box-type burner is therefore used.

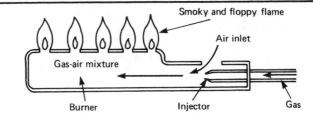

Fig 1 Gas pressure too low or injector bore too large

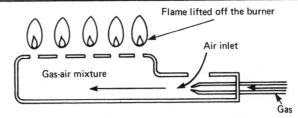

Fig 2 Gas pressure and injector bore correct but with no retention flame

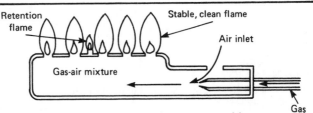

Fig 3 Gas pressure and injector bore correct with a retention flame

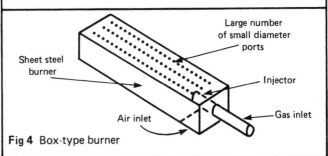

Fig 4 Box-type burner

127

CONVENTIONAL GAS FLUE

Notes

1   When it is necessary to site a gas appliance away from the outside wall a conventional flue is required. This should be built either of precast concrete blocks or of pipes made from stainless steel, enamelled steel, cast iron or asbestos-cement.

2   The flue has four parts
(a) primary flue
(b) draught diverter
(c) secondary flue or flue proper
(d) terminal.

3   It is important that the flue is correctly sized and the factors which determine the flue size are
(a) heat input of the appliance;
(b) resistance to flow of products of combustion by bends junctions and terminals;
(c) length of flue.

4   The pipe sockets should face upwards and be filled with fire cement. The number of bends in the pipe reduced to a minimum.

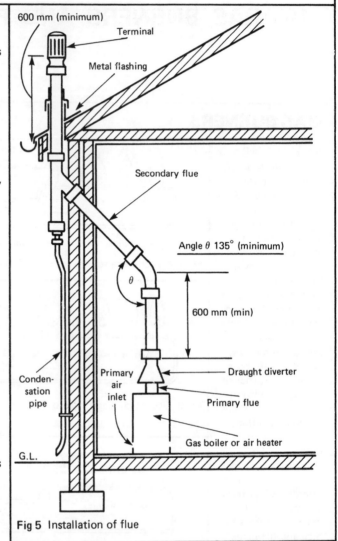

Fig 5 Installation of flue

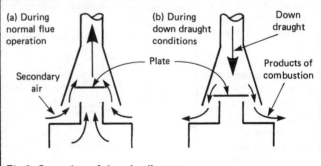

Fig 6 Operation of draught diverter

SHARED FLUES — Se-DUCT

Notes

1   In order to save on the cost of providing gas flues in multi-storey buildings (which at one time required a separate flue for each gas appliance) the device now known as the Se-duct was developed.

2   The Se-duct system consists of a vertical precast concrete duct open at each end to the atmosphere. Room sealed gas appliances (which must be provided with a flame failure safety device) are connected to the duct and these discharge their products of combustion back into the duct.

3   The degree of dilution in the duct must be sufficient to provide a maximum of 2%, by volume, of CO_2 at the top appliance.

4   Building Regulations give the sizes of flues for varying heat inputs.

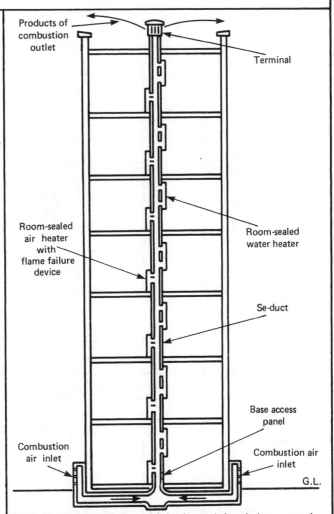

Products of combustion outlet

Terminal

Room-sealed air heater with flame failure device

Room-sealed water heater

Se-duct

Base access panel

Combustion air inlet

Combustion air inlet

G.L.

Fig 7 Typical installation with horizontal duct below ground

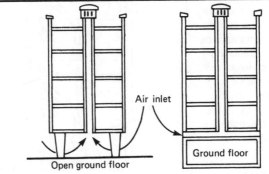

Air inlet

Open ground floor

Ground floor

Fig 8 Installation with an open ground floor

Fig 9 Installation with a horizontal duct in the ground floor ceiling

SHARED FLUES — U-DUCT AND SHUNT-DUCT

Notes

1   The U-duct system offers an alternative to the Se-duct in cases where there are difficulties in arranging the air supplies at the bottom of the duct. The U-duct has the advantages of the Se-duct system but requires two vertical ducts which in total area are greater than the Se-duct.

2   The down flow duct is the combustion air inlet duct which draws air from the roof level. The appliances of the room sealed type, fitted with a flame failure safety device, are connected to the upflow duct.

3   Stable duct flow under all wind conditions is achieved by using a balanced flue terminal designed to provide identical inlet and outlet exposure.

4   The shunt duct system is for flueing of conventional type appliances and shows a great saving in space over providing each appliance with an individual flue. It is the most economical shared flue.

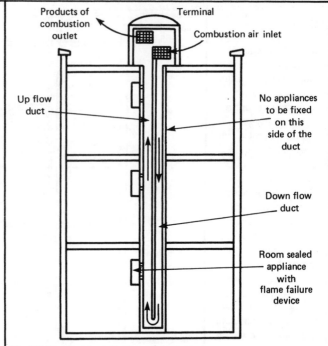

Fig 10 Typical installation of U-duct

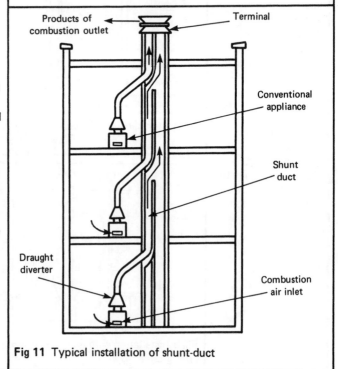

Fig 11 Typical installation of shunt-duct

FAN DILUTED GAS FLUES

Notes

1   Central plant installations usually require less expenditure of flues and if the boiler room is at roof level the boilers will require only short flues. Due to the increasing number of high-rise shops and offices, problems arise in the provision of flues for gas-fired boilers on the ground floor.

2   With this in mind, the fan-assisted flue with dilution air was developed. By means of a fan, fresh air is drawn into the flue, mixed with the products of combustion from the boilers and discharged to the external air.

3   In order to ensure that the boilers are fired when the fan is in operation a fan failure device is fitted to the flue. Failure of the fan switches off the boilers.

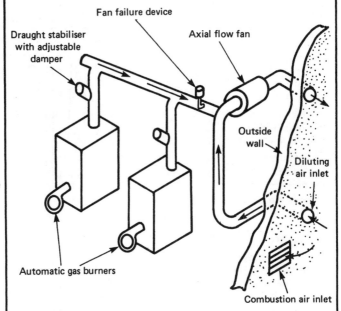

Fig 12 Installation using one outside wall and boilers with automatic burners

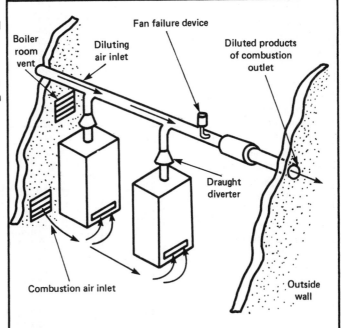

Fig 13 Installation, using two outside walls and boilers with draught diverters

131

BALANCED FLUE GAS APPLIANCES

Notes

1   The balanced flue appliance has the air inlet and flue outlet (except for the purpose of lighting) sealed from the room in which it is installed. It is more efficient than the conventional flue appliance and there is less risk of the products of combustion entering the room.

2   The balanced flue appliance is designed to draw in the air required for combustion from a point immediately adjacent to where it discharges its products of combustion. These inlet and outlet points must be inside a windproof terminal sited outside the room in which the appliance is installed.

3   The appliance does not require a long flue and there is less risk of the products of combustion entering the room. The appliance must always be fitted to an outside wall.

4   A gas appliance in a bath or shower room or in a garage must be a balanced flue appliance.

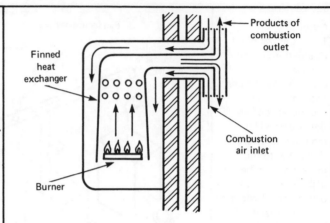

Fig 14 Balanced flue water heater

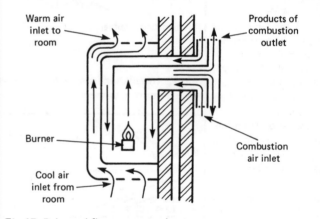

Fig 15 Balanced flue convector heater

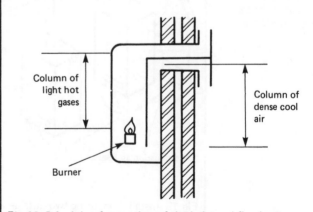

Fig 16 Principle of operation of the balanced flue heater

VENTILATION OF GAS APPLIANCES

Notes

1   Room sealed appliances installed in a compartment or in an enclosure such as a cupboard do not require an air vent for combustion but air vents are necessary to remove excess heat.

2   Ventilation can be either by natural means or by means of a fan to increase the rate of air flow into the compartment.

3   With open or conventionally flued appliances, provision must be made for the introduction of combustion air into the room or compartment. A flueless appliance requires an air vent direct to the external air, plus an openable window in the room or internal space in which it is installed.

4   Vents should be sited where they cannot be easily blocked or flooded. Vents at high level should be as close as possible to the ceiling. Low level vents should be located at not more than 450 mm above floor level.

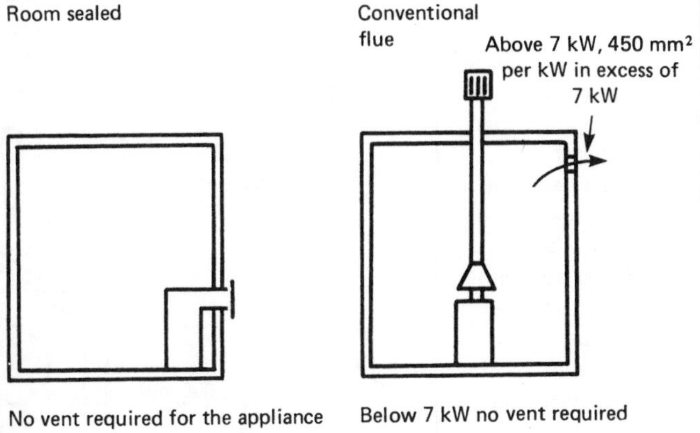

Room sealed

Conventional flue

Above 7 kW, 450 mm² per kW in excess of 7 kW

No vent required for the appliance

Below 7 kW no vent required

Fig 17 In a room

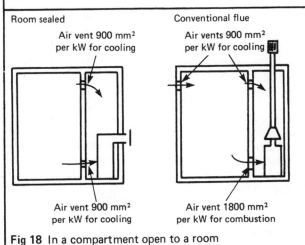

Room sealed

Air vent 900 mm² per kW for cooling

Air vent 900 mm² per kW for cooling

Conventional flue

Air vents 900 mm² per kW for cooling

Air vent 1800 mm² per kW for combustion

Fig 18 In a compartment open to a room

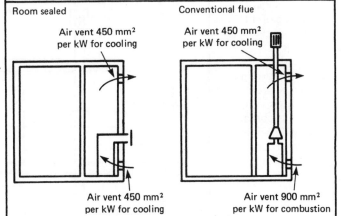

Room sealed

Air vent 450 mm² per kW for cooling

Air vent 450 mm² per kW for cooling

Conventional flue

Air vent 450 mm² per kW for cooling

Air vent 900 mm² per kW for combustion

Fig 19 In a compartment open to the outside

POSITIONS OF GAS FLUE TERMINALS

Notes

1 Wall termination for open or conventionally flued appliances is not recommended. The Building Regulations require that air must pass across a terminal at all times and the outlet must be at least 300 mm away from wall opening.

2 A terminal must not be placed in an air pressure zone. These zones are: immediately below the eaves or a balcony, in the corner of a building or close to rainwater or soil pipes.

3 Where a terminal is within 600 mm below a plastic gutter an aluminium shield 1.500 long should be fitted to the underside of the gutter immediately above the terminal.

4 Care must be taken to prevent re-entry of the products of combustion.

5 The terminal should be protected with a guard if it could come into contact with people or be subject to damage.

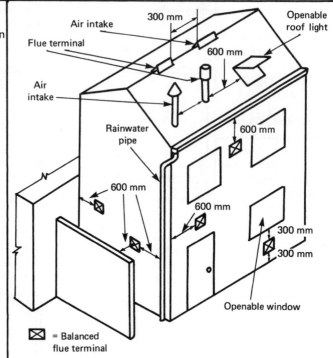

⊠ = Balanced flue terminal

Fig 20 Positions of terminals on pitched roof and external walls (minimum distances)

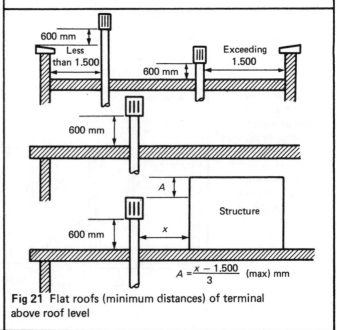

$$A = \frac{x - 1.500}{3} \text{ (max) mm}$$

Fig 21 Flat roofs (minimum distances) of terminal above roof level

17 OIL FIRING

OIL-FIRED BURNERS

Notes

1   There are two types of oil burner:
(a) vapourising;
(b) atomising:
   The natural draught vapourising burner consists of a cylindrical pot which is fed with oil at its base from a constant oil level controller.

2   When the burner is lighted, a thin film of oil burns in the bottom, heat is generated and the oil is vapourised. When this vapour comes into contact with air entering the lowest holes it mixes with the air and ignites. At full firing rate more air and oil mix until a flame burns out of the top of the burner.

3   The pressure jet atomising burner has an atomising nozzle. This produces a fine spray of oil which is mixed with air forced into the burner by a fan. Ignition electrodes produce a spark and ignites this air/oil mixture.

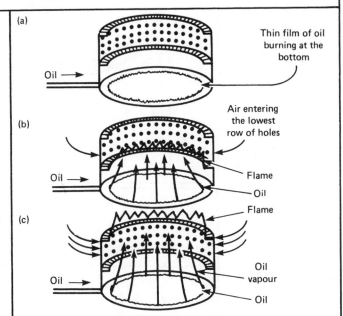

(a)

Thin film of oil burning at the bottom

Oil →

(b)

Air entering the lowest row of holes

Oil →

Flame
Oil
Flame

(c)

Oil →

Oil vapour
Oil

Fig 1 Natural draught pot vapourising burner

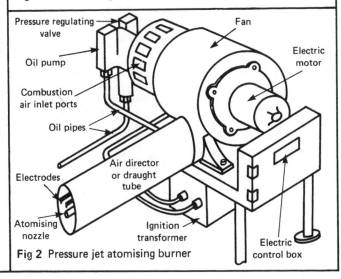

Pressure regulating valve
Oil pump
Combustion air inlet ports
Oil pipes
Electrodes
Atomising nozzle
Air director or draught tube
Ignition transformer
Fan
Electric motor
Electric control box

Fig 2 Pressure jet atomising burner

WALLFLAME OIL BURNER — OIL-LEVEL CONTROLLER

Notes

1 The wall flame burner consists of a steel base plate securing a centrally placed electric motor. The armature of this motor is wound on a hollow metal shroud which dips into an oil well. A constant oil-level controller feeds the well, just covering the edge of the shroud. The shroud is circular with its internal diameter increasing towards the top from which two holes connect with a pair of oil pipes.

2 When the motor is switched on, oil is drawn up to the pipes and is thrown onto the flame ring. At the same time air is forced onto the rings by the fan.

3 This air/oil mixture is ignited by the electrodes.

4 The constant oil-level controller is used to feed vapourising burners.

5 If the inlet valve fails to close, oil flows into the trip chamber. The trip float rises and operates the trip mechanism thus closing the valve.

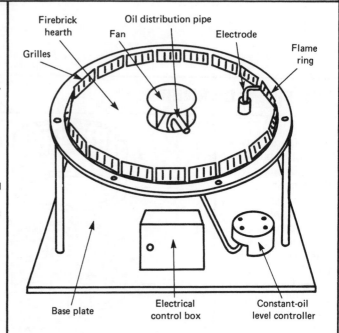

Fig 3 Wallflame rotary vapourising burner

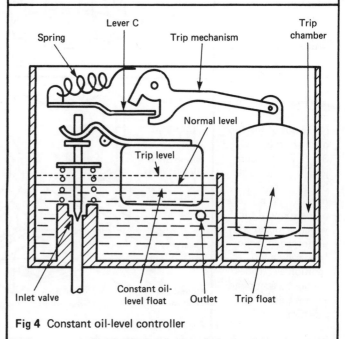

Fig 4 Constant oil-level controller

INSTALLATION OF OIL TANK AND OIL SUPPLY

Notes

1   An oil storage tank is usually rectangular with a top designed to shed water. For domestic work the capacity of the tank is 2275 litres. The tank should be made from ungalvanised welded carbon steel or sectional pressed ungalvanised carbon steel. Internal strutting may be required to resist bursting when full.

2   Unless the tank is underground, a fire burner provided or the wall of the building is protected the tank must be sited at least 1.800 from the building. The tank must also be within 30 m of the oil tanker vehicle access point (unless an extended fill line is provided).

3   Fully annealed copper tubing or steel pipe may be used for the oil pipeline. The head of oil from the outlet pipe to the centre of the burner should be at least 300 mm. The maximum head above the burner is 3 m.

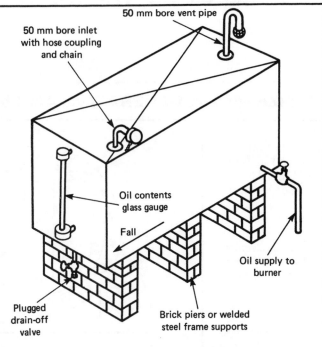

50 mm bore vent pipe

50 mm bore inlet with hose coupling and chain

Oil contents glass gauge

Fall

Oil supply to burner

Plugged drain-off valve

Brick piers or welded steel frame supports

Fig 5 Installation of outside oil storage tank

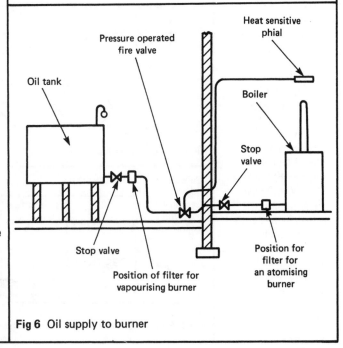

Heat sensitive phial

Pressure operated fire valve

Oil tank

Boiler

Stop valve

Stop valve

Position for filter for an atomising burner

Position of filter for vapourising burner

Fig 6 Oil supply to burner

18 ELECTRIC WIRING SYSTEMS

THREE-PHASE GENERATION AND SUPPLY

Notes

1   In 1831 Michael Faraday plunged a bar magnet into a coil of wire and thus produced electricity. This method of producing electricity is used today but the coils of wire are cut by a magnetic field as the magnet rotates. These coils of wire (or stator windings) have an angular space of 120°and the voltages produced are out of phase by this angle for every revolution of the magnets, thus producing a three-phase supply.

2   A three-phase supply provides 73% more power than single phase supply for the addition of a wire.

3   With three-phase supply, the voltage between the two line cables is 1.73 times that between the neutral and any one of the line cables.

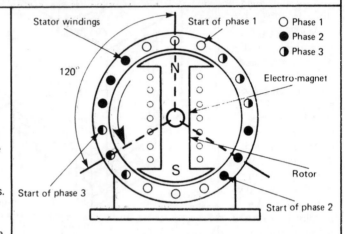

Fig 1 Simplified detail of three-phase generator or alternator

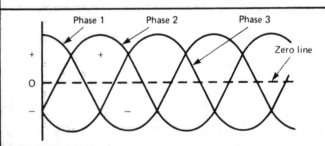

Fig 2 Three-phase supply

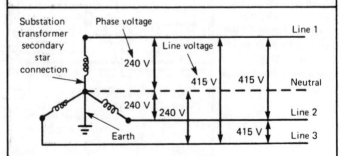

Fig 3 Relationship between line and phase voltage

ELECTRICITY SUPPLY FROM THE 'GRID'

Notes

1   Electricity in the UK is produced at the generating stations at 25 kV, three phase, 50 Hertz (Hz) and is fed through transformers which step up the voltage to 132, 275 or 400 kV before being connected to the National grid.

2   Power to the larger towns is taken by cable or overhead line at 132 kV or 33 kV where it is transformed to 11 kV for supply by under-ground cable to sub-stations. Here the power is converted yet again, to 415 kV and 240 V.

3   The supply to houses and other small buildings is by an underground ring circuit from the sub-stations. Supplies to factories and other large buildings are taken from the 132 V or 33 kV main supply. These larger buildings require their own transformer.

4   A transformer usually has delta-star connections to provide a four wire supply to the building.

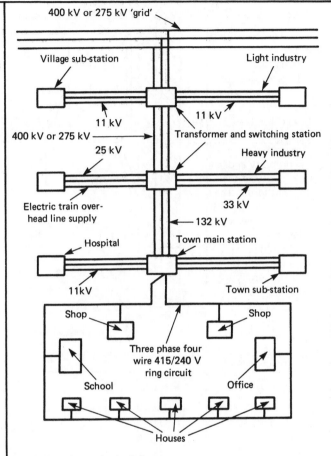

Fig 4 Supply to the buildings

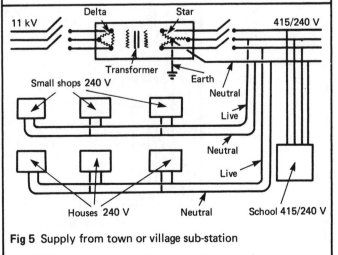

Fig 5 Supply from town or village sub-station

139

ELECTRICITY INTAKE TO BUILDING

Notes

1   The termination and metering of service cables to buildings is affected by arrangement with the supply authority for the district. All equipment up to and including the meter is the property and responsibility of the supply authority. The supply authority must provide a fusible cut-out neutral link, a meter and in some cases a transformer.

2   The supply to small buildings is normally brought into the buildings through a trench and left in a position near the entrance ready for the installation of meters and fusegear. Underneath roadways the cable should be at a depth of at least 760 mm and at least 460 mm below open ground.

3   The meter may be sited inside the building or outside in a glass reinforced plastic meter cabinet.

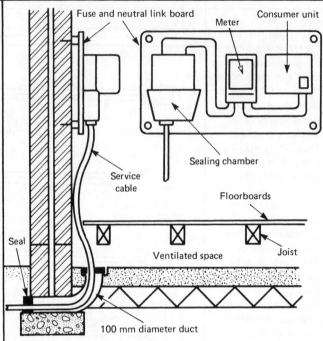

Fig 6 Underground service entry

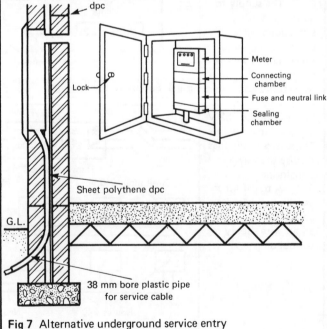

Fig 7 Alternative underground service entry using external meter cabinet

PRIVATE SUBSTATION — TRANSFORMER

Notes

1 A substation is required for the conversion, transformation and control of electric power. When a building requires more power than the local low or medium voltage (240 V or 415 V) distribution system can provide, a substation must be built on the customer's premises. This is fed by high voltage cables from the Electricity Board's nearest switching station.

2 The requirements of a substation depend upon the number and size of the transformers and switches.

3 A transformer consists basically of two electric windings magnetically interlinked by an iron core. An alternating electro-motive force applied to one of the windings produces, by electro-magnetic induction, a corresponding electro-motive force in the other winding.

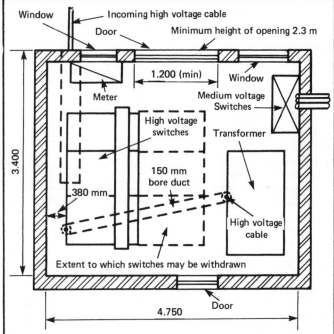

Fig 8 Construction and layout of substation

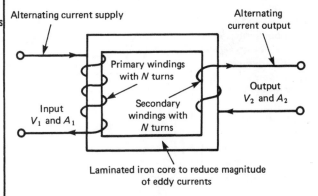

If losses are ignored, the following relationships of a transformer apply

$$\frac{V_1}{V_2} = \frac{N_1}{N_2} = \frac{I_2}{I_1}$$

Where V_1 = primary voltage I_1 = primary current
        V_2 = secondary voltage I_2 = secondary current
        N_1 = number of primary turns
        N_2 = number of secondary turns

Fig 9 Principle of transformer

BONDING OF SERVICES — RING CIRCUIT

Notes

1   The IEE Regulations require that the metal sheaths and armour of all cables operating at low or medium voltage, along with all metal conduits, ducts, trunking and bare earth continuity conductors of other fixed metal work shall be effectively segregated or earth bonded.

2   The bonding of the services shall be as close as possible to the point of entry of the services into the building.

3   A ring circuit is used for single-phase power and consists of two current-carrying conductors and an earth looped into each socket outlet. In a domestic building a ring circuit may serve an unlimited number of socket outlets provided that there is one ring for each 100m² of floor area. Plug outlets on the ring are provided with 13 A or 3 A fuses to suit the appliance connected to the plug. The number of socket outlets from a spur shall not exceed the number of socket outlets and fixed appliances on the ring.

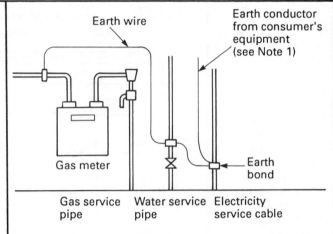

Fig 10 Bonding of services at intake

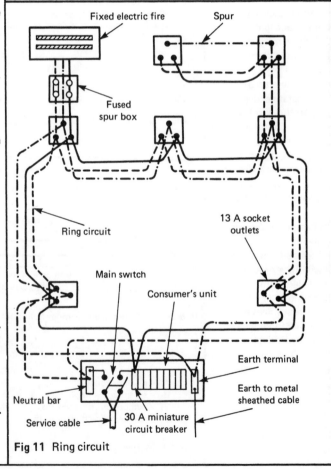

Fig 11 Ring circuit

LIGHTING CIRCUITS — 1

Notes

1 With the one-way switch circuit the single-pole switch must be connected to the live conductor. To ensure that both live and neutral conductors are isolated from the supply a double pole switch may be used. Provided that the voltage drop is not exceeded, two more more lamps may be controlled by a one-way single pole switch.

2 In principle the two-way switch is a single pole change-over switch and interconnected in pairs; two switches provide control of the lamp from two positions. Two-way switching may be used on landings, halls, bedrooms and corridors.

3 In large buildings, every access point should have its own lighting control switch and any number of these may be incorporated into a two-way switch circuit. These additional controls are known as intermediate switches.

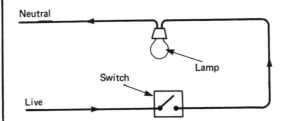

Fig 12 One-way single pole switch circuit controlling one lamp.

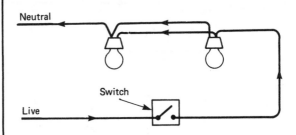

Fig 13 One-way single pole switch circuit controlling two or more lamps

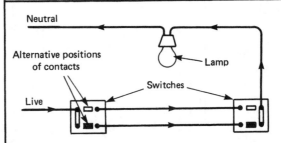

Fig 14 Two-way switching

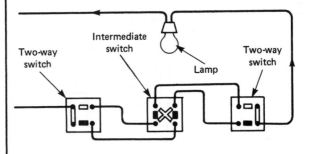

Fig 15 Two-way switching with one intermediate switch

LIGHTING CIRCUITS — 2

Notes

1   The purpose of a 'master' switch is to limit or vary the scope of control afforded by other switches in the same circuit. If a 'master' switch (possibly one with a detachable key operation) is fixed near the main door of a house or flat the householder is provided with a means of controlling all the lamps from one position. This is useful when going out or returning during the hours of darkness.

2   A sub-circuit for lighting is generally limited to a total load of 100 W and requires a 5 A miniature circuit breaker. In large buildings however 15 A miniature circuit breakers are often used due to the higher load. Wiring for lighting is normally carried out on what is known as the 'looping-in' system.

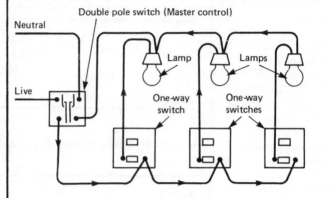

Fig 16 'Master' control wiring circuit

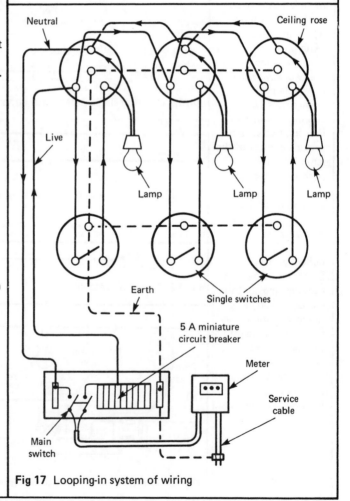

Fig 17 Looping-in system of wiring

CURRENT-OPERATED EARTH LEAKAGE CIRCUIT BREAKERS

Notes

1   In the single-phase, current-operated earth leakage circuit breaker the load current is fed through two equal and opposing coils, wound on a common transformer core. When the live and neutral currents are balanced, as they should be in a normal circuit, they produce equal and opposing fluxes in the transformer or magnetic core. This means that no resultant voltage is generated in the fault detector coil. If, however, due to an earth fault, more current flows in the live coil than in the neutral coil, an out-of-balance flux will be produced which will be detected by the fault detector coil. The current in the fault detector coil is arranged to trip a circuit breaker.

2   A three-phase earth leakage circuit breaker operates on the same principle as the single-phase type but there are three equal and opposing coils.

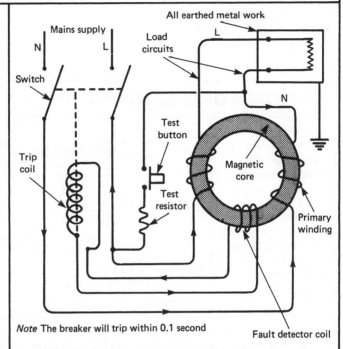

Note The breaker will trip within 0.1 second

Fig 18 Single-phase earth leakage circuit breaker

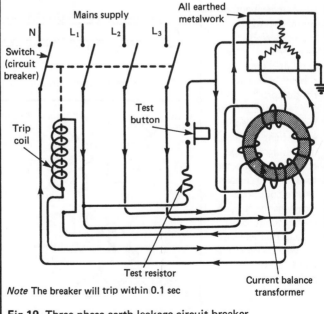

Note The breaker will trip within 0.1 sec

Fig 19 Three-phase earth leakage circuit breaker

145

19 METHODS OF ELECTRIC WIRING

METHODS OF ELECTRIC WIRING — 1

Notes

1   Armoured cable is used for mains and submains. The cable is laid below ground level but is brought above ground at substations and inside buildings. High voltage cable is protected below ground by tiles.

2   A conduit is either metal or PVC tube into which insulated cables are drawn. The function of a conduit is to provide mechanical protection of the cables, permit rewiring and, if the conduit is metal, to provide an earth conductor.

3   Steel conduit may be obtained in either light or heavy gauge. Light gauge conduit is connected to fittings by means of grip couplings. Heavy gauge conduit is connected to fittings by means of screwed threads. The conduit must be protected from corrosion.

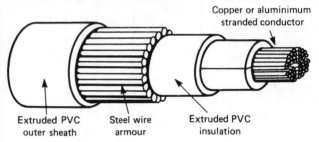

Copper or aluminimum stranded conductor

Extruded PVC outer sheath   Steel wire armour   Extruded PVC insulation

Fig 1 Armoured three phase four wire cable for laying below ground level

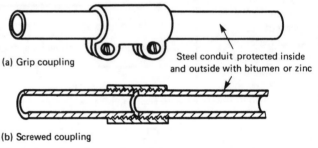

(a) Grip coupling

Steel conduit protected inside and outside with bitumen or zinc

(b) Screwed coupling

Fig 2 Couplings for steel conduit

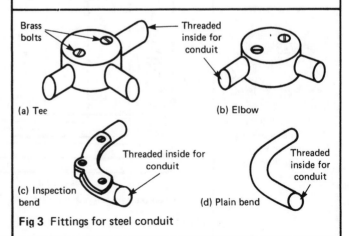

Brass bolts   Threaded inside for conduit

(a) Tee   (b) Elbow

Threaded inside for conduit

(c) Inspection bend

Threaded inside for conduit

(d) Plain bend

Fig 3 Fittings for steel conduit

METHODS OF ELECTRIC WIRING — 2

Notes
1 In the manufacture of mineral insulated cable, copper or aluminium rods are introduced into a tube of suitable diameter. This tube is then filled with magnesium oxide powder which has been subjected to controlled drying. It is essential when installing the cable that the magnesium oxide powder (which is hygroscopic) does not come into contact with the surrounding air. The connections to joint boxes must be sealed. The cable provides an excellent earth conductor and is the safest form of installation; it also resists corrosion and is unaffected by extremes of heat.
2 PVC or rubber insulated cable is cheap and quick to install. The cable however, does not provide much protection from mechanical damage or heat.

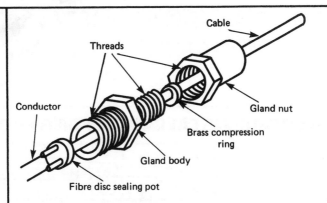

Fig 4 Exploded view of termination joint for mineral insulated copper or aluminium covered cable

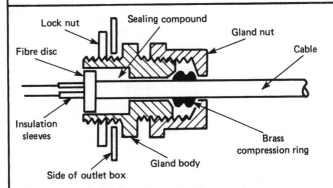

Fig 5 Section of termination joint for mineral insulated copper or aluminium covered cable (MICC or MIAC cable)

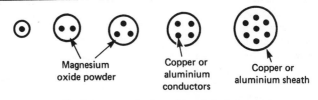

Fig 6 Core arrangements of mineral insulated copper or aluminium covered cables

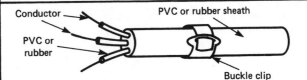

Fig 7 PVC or rubber insulated cable

20 ELECTRICAL INSTALLATIONS FOR LARGE BUILDINGS

ELECTRIC INSTALLATION FOR A FACTORY

Notes

1   If the electrical load is not too high, a factory may be supplied with a 3-phase, 4-wire 415 V service cable. For the supply to three phase motors an overhead busbar system is often used with supplies to the motors through steel conduit via push button switches. The switches must be within easy reach of the operator and contain a device to prevent restarting of the motor after stopping due to a power failure.

2   The overhead busbar system provides an easy means of reconnecting a motor after it has been moved to another position.

3   For lighting and single-phase power circuits a separate distribution fuseboard is required.

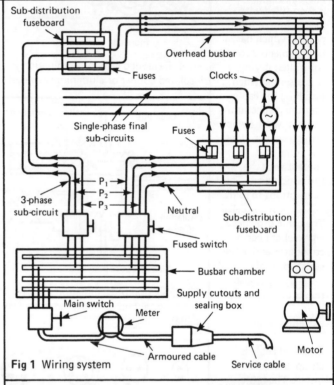

Fig 1 Wiring system

Labels: Sub-distribution fuseboard; Overhead busbar; Fuses; Clocks; Single-phase final sub-circuits; Fuses; P_1 P_2 P_3; 3-phase sub-circuit; Neutral; Sub-distribution fuseboard; Fused switch; Busbar chamber; Supply cutouts and sealing box; Main switch; Meter; Motor; Armoured cable; Service cable

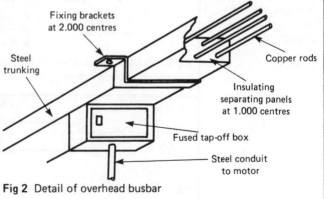

Fig 2 Detail of overhead busbar

Labels: Fixing brackets at 2.000 centres; Steel trunking; Copper rods; Insulating separating panels at 1.000 centres; Fused tap-off box; Steel conduit to motor

ELECTRIC SUPPLY TO GROUPS OF LARGE BUILDINGS

Notes

1 For large development schemes with several buildings, either a radial or a ring distribution may be used for the electric supplies.

2 In the radial system separate underground cables are laid from the intake room to each building. The system uses more cable than the ring system but only one fused switch is required below the distribution boards in each building.

3 In the ring-circuit system an underground cable is laid starting from the substation and looping-in to each building. In order to isolate the supply, two fused switches are required below the distribution boards in each building. Current flows in both directions from the intake room and this provides a better balanced supply than the radial system. If the cable on the ring is damaged at any point the cable may be isolated for repair without the loss of supply to any one of the buildings.

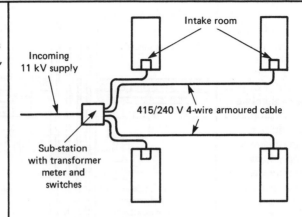

Fig 3 Radial distribution (block plan)

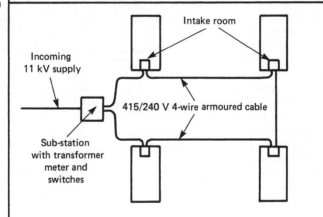

Fig 4 Ring distribution (block plan)

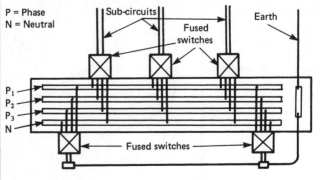

Fig 5 Detail of equipment in the intake room for the ring distribution

149

RISING MAIN ELECTRIC DISTRIBUTION SYSTEM

Notes

1   The rising main distribution system is used for electrical supplies to multi-storey flats or offices. Copper or aluminium busbars are supported by insulated bars across the chamber. The supply to each floor is connected to the rising main by means of tap-off units. In order to balance the electrical load between the phases, the connections to each floor should be spread between the phase bars. In a six-storey building having the same loading on each floor, two floors would be supplied from separate phases. Blocks of flats will require meters on each tap-off unit.

2   To prevent the spread of fire and smoke, fire barriers are fixed at each floor level and the chamber must be fire stopped to the full depth of the floor.

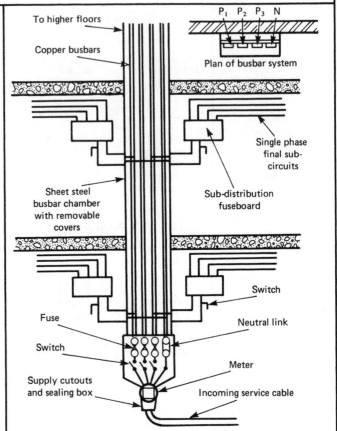

Fig 6 Detail of rising main system

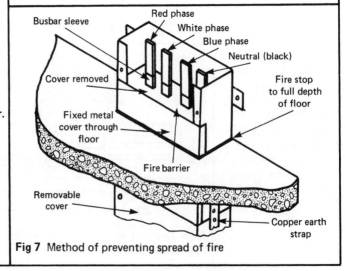

Fig 7 Method of preventing spread of fire

FLOOR AND SKIRTING DUCTS

Notes

1   For open plan offices a grid layout of the floor ducts may be used. This method permits outlets for both telephone and power supplies to be taken directly from a junction box to the nearest desk. In a large room, this is the only practical way for desks in the middle of the room to be served without exposed lengths of cable or telephone cords.

2   For partitioned offices or similar buildings a branching duct layout may be used. The branches can either terminate at junction boxes near to the wall or extend to wall outlets.

3   Where an electrical supply is to be run alongside telephone cables in a combined duct it is essential that segregation of the services is provided.

4   For some buildings a metal skirting duct may be used with outlets provided.

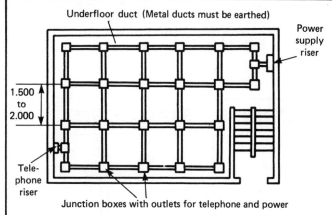

Fig 8 Grid layout floor duct

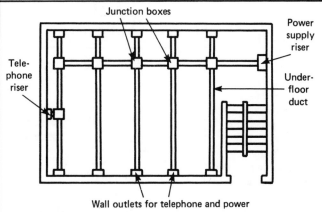

Fig 9 Branching layout floor duct

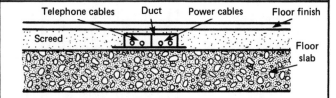

Fig 10 Section through floor duct

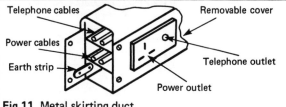

Fig 11 Metal skirting duct

151

21 TELEPHONES AND ELECTRIC HEATING

TELEPHONE INSTALLATIONS

Notes
1   The external telephone cables may be either overhead or below ground. The modern tendency is to provide a below ground cable system. The cable from the below ground system is taken in to the building below pavement level and the point of entry should be decided with the telephone manager.
2   In large buildings the incoming cable supplies a main distribution unit to which distribution cables are connected to supply various parts of the building. The cables supply both switchboards and individual telephones. There is a limit to the number of cables which can be taken from a riser to the telephones and the Telephone Manager must be consulted about this.

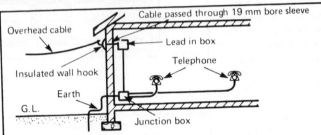

Fig 1 Overhead telephone cables

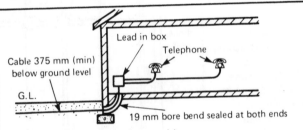

Fig 2 Underground telephone cable

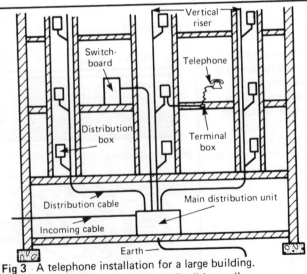

Fig 3 A telephone installation for a large building. Cables inside the building (not the flexible cord) must be concealed in ducts and the system earthed.

ELECTRIC SPACE HEATING — 1

Notes

1   Because electricity from an alternating current supply cannot be stored, the Area Boards encourage the use of 'off peak' heating and offer cheaper tariffs for this type of load. A white meter system controls the 'off peak' loads. Underfloor heating makes use of the thermal storage properties of the concrete floor for storing the heat generated by the electric currents.

2   Block storage heaters are rated between 1kW and 6 kW and make use of concrete blocks to store the heat produced by the electric currents. Heaters with a fan give greater control of the heat output.

3   Electrically-heated ceilings use 'on-peak' electricity. The heating element is flexible glasscloth with a conducting silicone elastomer.

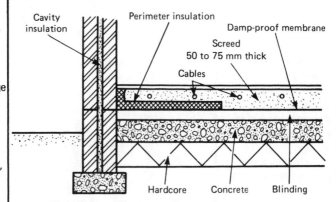

Fig 4 Section through solid ground floor with heating cables

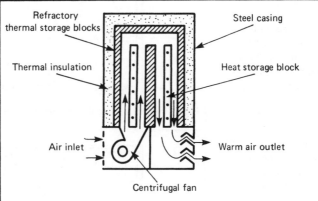

Fig 5 Block storage heater with fan

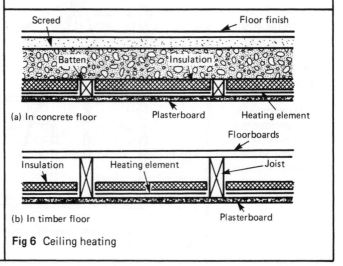

(a) In concrete floor

(b) In timber floor

Fig 6 Ceiling heating

ELECTRIC SPACE HEATING — 2

Notes

1 Electrically-heated warm air systems, supplied from a central unit, are a development from the 'off peak' block storage heaters with a fan (see Space Heating — 1). The input loadings are from 6 kW to 12 kW. A room thermostat keeps the air temperature at the desired level by operating the fan. The warm air is conveyed by insulated ducts and can be directed into any rooms by opening the appropriate supply register.

2 In the 'stub' duct system the unit is installed in a central position and warm air is carried through into the rooms by short ducts.

3 In the 'radial' duct system warm air from the unit feeds through a number of radial ducts which supply warm air to the outlet registers. This system warms the cooler perimeter of the rooms and increases thermal comfort.

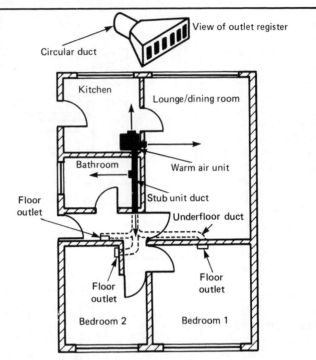

Fig 7 Plan of a bungalow showing a 'stub' duct warm air system

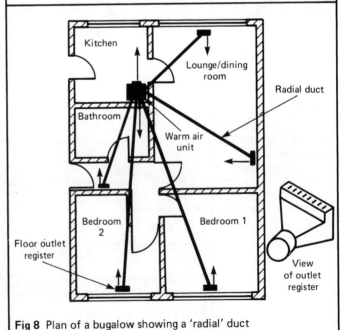

Fig 8 Plan of a bugalow showing a 'radial' duct warm air system

ELECTRIC SPACE HEATING — 3

Notes

1 The heat output from panel heaters is mainly radiant. The panel surface has a surface operating temperature of between 204°C and 240°C.

2 The infra-red heater consists of an inconel-sheathed element or a nickel chrome spiral element in a glass tube backed by a reflector.

3 Oil-filled heaters are similar in appearance to pressed steel hot water radiators, they give radiant and convected heat.

4 The convector heater usually has two elements under independent control. It is used where a constant level of background warmth is required.

5 A parabolic reflector fire has the heating element in the focal point to give efficient radiant heating.

6 Wall mounted fan heaters may have a two-speed fan which is quiet in operation.

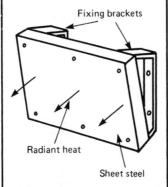

Fig 9 Wall mounted radiant panel heater

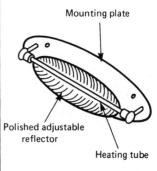

Fig 10 Wall mounted infra-red heater

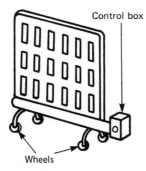

Fig 11 Oil-filled portable heater

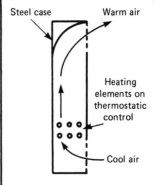

Fig 12 Convector heater

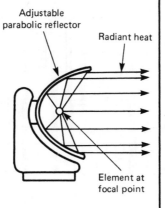

Fig 13 Portable parabolic reflector fire

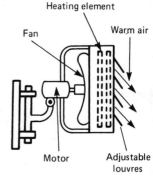

Fig 14 Wall mounted fan heater

ELECTRIC SPACE HEATING — 4

Notes

1   As an alternative to storing, the heat produced by the electric currents in concrete floors or storage heaters by 'off-peak' electricity may be converted to heat energy and used to heat water in a vertical or a horizontal cylinder. This method is known as hot water thermal storage heating. It is easier to provide thermostatic control with this system than with the underfloor, or block storage heating systems.

2   In this system the water is heated up to 180°C and circulated by a pump through heat emitters in the building. There is sufficient hot water in the cylinder to heat the building during the daytime. During the off peak period, usually between 1900 and 0700 hours, the water is reheated.

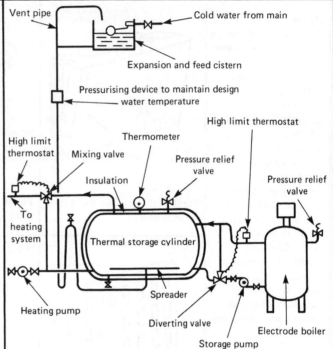

Fig 15 Heating system using water

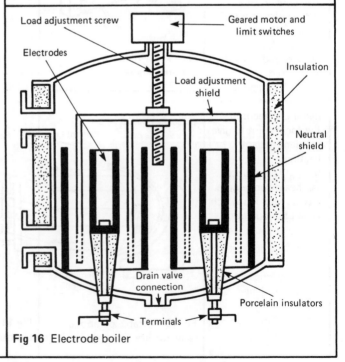

Fig 16 Electrode boiler

22 CONSTRUCTION SITE ELECTRICITY

CONSTRUCTION SITE ELECTRICITY

Notes

1   Electricity for building sites may be obtained from portable generators or the local area electricity board. The most convenient being from the area electricity board. To ensure that the supply is available when needed a written application for a temporary supply of electricity is made as soon as possible.

2   Before the supply is used it is essential that the installation is inspected and tested by a competent electrician. Inspections and tests must also be carried out during the construction.

3   For site lighting and portable tools a 110 V supply must be used. Good earthing is essential.

4   Fig 2 shows methods to prevent contractors' plant fouling overhead power lines. The location of underground cables must be established.

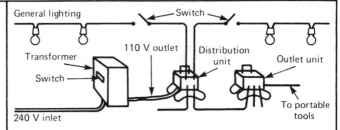

Fig 1 Reduced voltage distribution

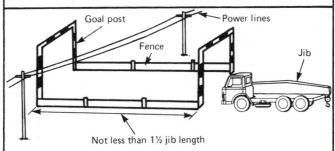

Fig 2 Goal posts (or barrier fences) give protection against contact with overhead power lines

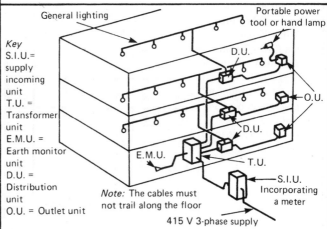

Key
S.I.U. = supply incoming unit
T.U. = Transformer unit
E.M.U. = Earth monitor unit
D.U. = Distribution unit
O.U. = Outlet unit

Note: The cables must not trail along the floor

Fig 3 Typical arrangement of distribution units and equipment

23 ELECTRIC AND SUPPLEMENTARY LIGHTING SYSTEMS

LIGHT AND LIGHT SOURCES

Notes

1   Light is a form of electro-magnetic radiation and is similar in nature and behaviour to radio waves at one end of the frequency spectrum and X-rays at the other.

2   Light is reflected from a polished (specular) surface at the same angle as it strikes it. A matt surface will reflect in a number of directions and a semi-matt surface will behave in a manner between a polished and a matt surface.

3   The unit of luminous intensity is a candela. The unit of light flux is the lumen (lm). The unit of illuminance is the lux (lx) and is the illumination produced by one lumen over an area of one square metre. Efficacy is the efficiency of lamps in 1m/w.

Angle of incidence θ_1 = Angle of reflection θ_2

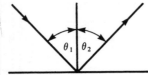

Fig 1 Light reflected from a polished surface

Light is reflected in all directions

Fig 2 Light reflected from a matt surface

Some light is scattered and some light is reflected directionally

Fig 3 Light scattered and reflected from a semi-matt surface

Light is scattered in all directions (diffusion)

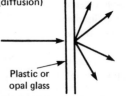

Plastic or opal glass

Fig 4 Light passing through a diffusing screen

Light is bent or refracted when passing through a surface between two media

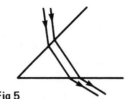

Fig 5

$$E = \frac{I}{d^2}$$

E = Illumination on surface in lux

I = Illumination intensity from source in candelas

d = distance from light source to surface in metres

Fig 6 Illumination produced from a light source perpendicular to the surface

$$E = \frac{I \cos \theta}{d^2}$$

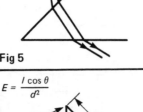

Source

Surface

Fig 7 Illumination produced from a light source not perpendicular to the surface

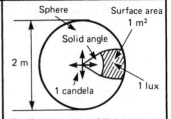

Sphere

Surface area 1 m²

Solid angle

2 m

1 candela

1 lux

Fig 8 Intensity of light and lux

ELECTRIC LAMPS

Notes

1   The tungsten iodine lamp is used in floodlighting. Evaporation from the filament is controlled by the presence of iodine vapour. The gas filled, general purpose filament lamp has a fine tungsten wire, sealed within a glass bulb. The wire is heated to incandescence (white heat) by the passage of an electric current.

2   Discharge lamps have no filament but produce light by excitation of a gas. When voltage is applied to the two electrodes ionisation occurs and when the voltage has reached a critical value, current flows between them. As the temperature rises, the mercury vapourises and discharge occurs between the main electrodes which causes light to be emitted.

3   A fluorescent tube is a low pressure mercury discharge lamp, in which the arc operates in a clear tube internally coated with fluorescent powder.

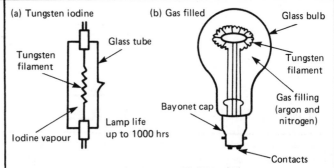

(a) Tungsten iodine — Glass tube, Tungsten filament, Iodine vapour, Lamp life up to 1000 hrs

(b) Gas filled — Glass bulb, Tungsten filament, Gas filling (argon and nitrogen), Bayonet cap, Contacts

Fig 9 Filament lamps (efficacy = 10-15 lm/W)

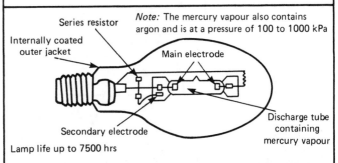

Note: The mercury vapour also contains argon and is at a pressure of 100 to 1000 kPa

Series resistor, Internally coated outer jacket, Main electrode, Secondary electrode, Discharge tube containing mercury vapour

Lamp life up to 7500 hrs

Fig 10 Mercury-vapour discharge lamp (efficacy = 50 lm/W)

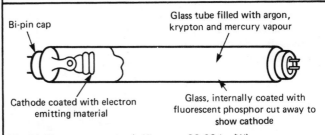

Bi-pin cap, Glass tube filled with argon, krypton and mercury vapour, Cathode coated with electron emitting material, Glass, internally coated with fluorescent phosphor cut away to show cathode

Fig 11 Fluorescent tube (efficacy = 20-60 lm/W)

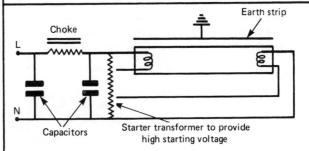

Choke, Earth strip, L, N, Capacitors, Starter transformer to provide high starting voltage

Fig 12 Control gear is needed to start the discharge and to keep the light steady during operation. A transformer provides a quick start.

LIGHTING FITTINGS — 1

Notes

1   Lighting fittings may be divided into three categories, namely:

(a) general utility fittings which are designed to give effective and economical illumination;

(b) special fittings usually with optical arrangements such as lenses or reflectors to give directional lighting;

(c) decorative fittings in which the light is there to be seen rather than to provide light to perform tasks.

2   From the optical consideration, the fitting should obscure the lamp from direct vision to reduce glare.

The British Zonal Method (BZ) divides light fittings into ten categories depending on the direction of light output. The direct fitting is efficient but the dull ceiling gives a feeling of gloom. Semi-direct and general diffusing fittings are popular for most installations.

3   Ventilated fittings allow the heat from the lamps to be drawn back for re-use and the light output is also improved.

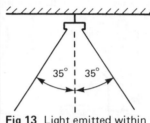

Fig 13 Light emitted within 35° of the vertical will not cause serious glare

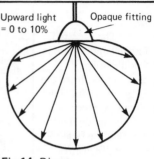

Fig 14 Direct

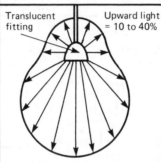

Fig 15 Semi-direct

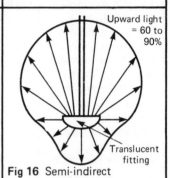

Fig 16 Semi-indirect

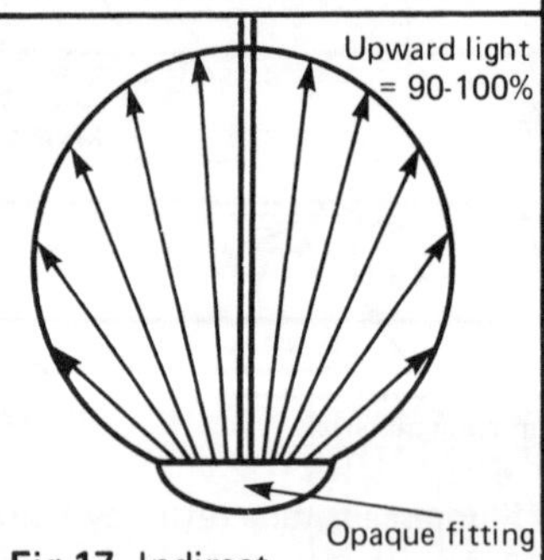

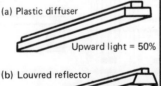

Fig 17 Indirect

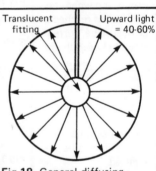

Fig 18 General diffusing

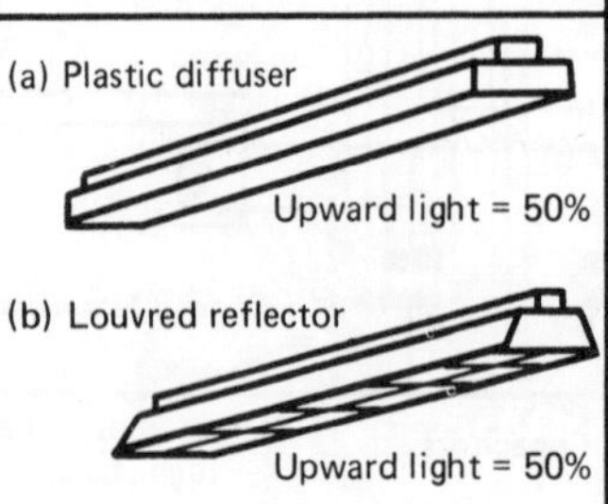

(a) Plastic diffuser
Upward light = 50%

(b) Louvred reflector
Upward light = 50%

Fig 19 Fittings used for fluorescent lamps

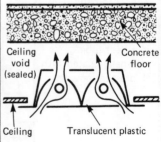

Fig 20 Ventilated fittings

LIGHTING FITTINGS AND LAMPS — 2

Notes

1   Fig 1 shows various types of fitting for industrial premises.

2   An illuminated ceiling provides cheerful lighting for offices and also good thermal insulation. Obstructions such as trunking will cast shadows on to the ceiling and must be avoided. The tubes must be mounted on batten fittings and the inside of the ceiling void painted white. In a dirty environment the ceilings may be made air- and dust-tight.

3   High pressure sodium discharge lamps give a consistent golden white light in which it is possible to distinguish colours. They are suitable for public, floodlighting, industrial and commercial lighting installations. Low pressure sodium discharge lamps produce light that is virtually monochromatic. The colour rendering is poor when compared to the high pressure lamp. The sodium vapour pressure for high and low pressure lamps is 0.5 Pa and 33 kPa respectively.

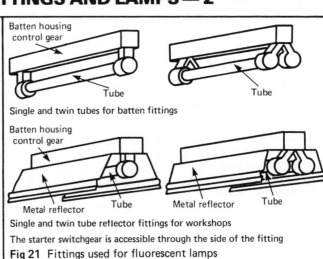

Single and twin tubes for batten fittings

Single and twin tube reflector fittings for workshops

The starter switchgear is accessible through the side of the fitting

Fig 21 Fittings used for fluorescent lamps

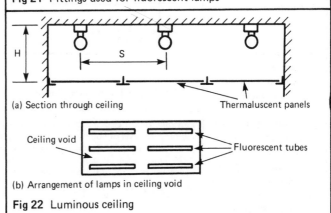

(a) Section through ceiling

(b) Arrangement of lamps in ceiling void

Fig 22 Luminous ceiling

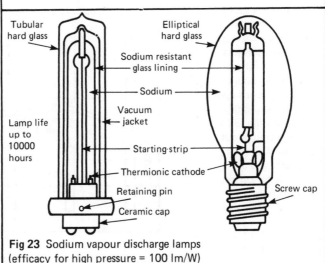

Fig 23 Sodium vapour discharge lamps
(efficacy for high pressure = 100 lm/W)

161

THE LUMEN METHOD OF LIGHTING DESIGN

Notes

1   The lumen method of design is the most widely used approach to the determination of lighting layout that will provide a service illumination on the working plane from lamps overhead in a substantially regular pattern. The method uses the formula shown in Fig 25 where

N = number of lamps;
E = lux on working plane;
A = floor area (m^2);
F = lumens per lamp;
U = utilisation factor;
M = maintenance factor;

The utilisation factor is the ratio of the lumens which arrive on the working plane to the total output of lamps in the scheme.

Maintenance factor takes into account the light lost due to dirt on the fittings and room surfaces.

2   The ratio of spacing to mounting height depends upon the British Zonal Classification. BZ3 fittings emit about 10% of their light upwards and are suitable for offices and schools. The ratio of spacing to mounting height is:
BZ1 and 2 = 1:1
BZ3 and 4 = 1.25:1
BZ5 to 10 = 1.5:1

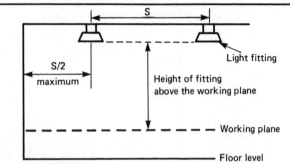

(a) Vertical section of a room

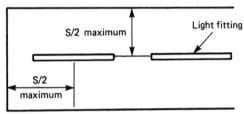

(b) Plan of a room

Fig 24 Method of spacing fluorescent tubes

Example. An office 8 m long by 7 m wide requires an illumination level of 400 lux on the working plane. It is proposed to use 80W fluorescent fittings having a rated output of 7375 lumens each. Assuming a utilisation factor of 0.5 and a maintenance factor of 0.8 design the lighting scheme.

$$N = \frac{E \times A}{F \times U \times M} \quad \therefore \quad N = \frac{400 \times 8 \times 7}{7375 \times 0.5 \times 0.8} \quad N = 7.59, \text{ use 8 fittings}$$

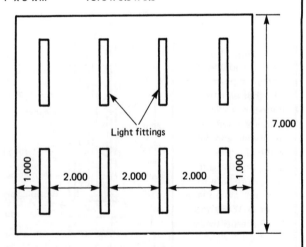

Fig 25 Layout of fluorescent tubes for the office

PERMANENT SUPPLEMENTARY LIGHTING OF INTERIORS

Notes

1   The first decision to be made before a building can be designed is how it is to be illuminated. Three types of illumination may be used namely:

A. daylighting alone, in which the window area occupies about 80% of the facades;

B. permanent supplementary artificial lighting of interiors in which the window area occupies about 20% of the facades

C. permanent artificial lighting of interiors in which there are no windows.

2   People usually prefer a view to the outside and therefore the choice of lighting for most building is in the form of type A or B. With type B the building may be wider because artificial lighting is used to supplement daylighting. Although the volume of the building is the same as for type A the building perimeter is less, thus saving in wall construction. The type B building also has lower heat gains and losses, less noise from outside and less window cleaning.

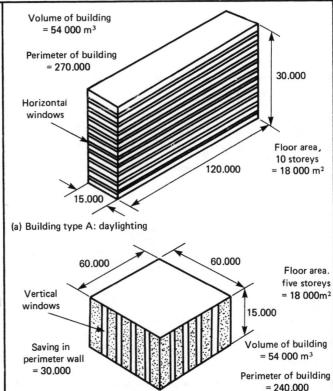

Volume of building = 54 000 m³

Perimeter of building = 270.000

Horizontal windows

30.000

Floor area, 10 storeys = 18 000 m²

120.000

15.000

(a) Building type A: daylighting

60.000   60.000

Vertical windows

Floor area. five storeys = 18 000m²

15.000

Saving in perimeter wall = 30.000

Volume of building = 54 000 m³

Perimeter of building = 240.000

(b) Building type B: permanent supplementary lighting

Fig 26 Elevations of alternative forms of building

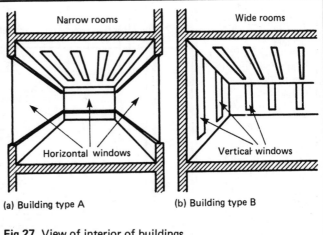

Narrow rooms

Wide rooms

Horizontal windows

Vertical windows

(a) Building type A

(b) Building type B

Fig 27 View of interior of buildings

163

24 DAYLIGHTING

DAYLIGHTING — 1

Notes

1   The daylight received inside a building can be expressed as 'the ratio of the illumination at the working point indoors to the total light available simultaneously outdoor'. This is usually expressed as a percentage and is known as the Daylight Factor. The factor includes light from

(a) *sky component*: light received directly from the sky; direct sunlight is excluded.

(b) *external reflected component*: light received from exterior reflecting surfaces

(c) *internal reflected component*: light received from internal reflecting surfaces.

2   If equal daylight factor contours are drawn for a room, it will show how daylight drops as one recedes from the window.

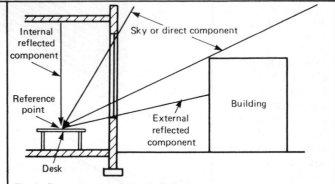

Fig 1 Components of 'daylight factor'

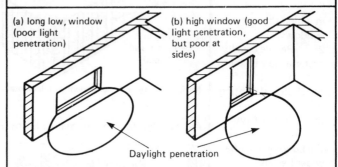

(a) long low, window (poor light penetration)

(b) high window (good light penetration, but poor at sides)

Daylight penetration

Fig 2 The effectiveness of window design

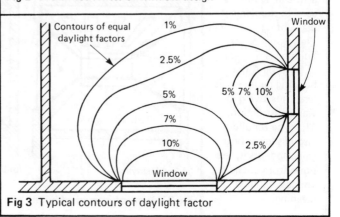

Contours of equal daylight factors

1%

2.5%

5%

7%

10%

Window

5% 7% 10%

2.5%

Window

Fig 3 Typical contours of daylight factor

DAYLIGHTING — 2

Notes

1 The effect of daylight in a room can be studied by use of scaled models and providing that surface textures and colours of the room surfaces are the same, an approximate result may be obtained.

2 An estimate of the effect of daylight in a room may also be made by the use of BRE daylight factor protractors and tables. The protractors are for determining, from a scaled drawing, the sky component from a sky of uniform luminance. There are five pairs for various types of windows.

3 Protractor No. 1 is placed on the cross section as shown. Readings are taken where the sight lines intesect the protractor scale. In the diagram the sky component = 8.5 - 4 = 4.5% and the angle of altitude 30°. The sky component of 4.5% must be corrected by use of protractor No. 2; which is placed on the plan as shown. Reading from protractor No. 2 are 0.25 and 0.1 and the correction factor is 0.25 + 0.1 = 0.35.

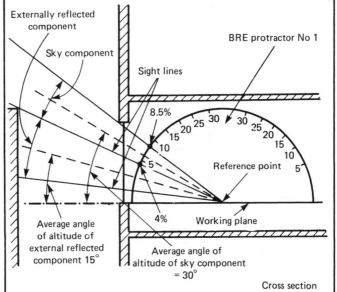

Fig 4 Use of BRE protractor No 1 (vertical windows)

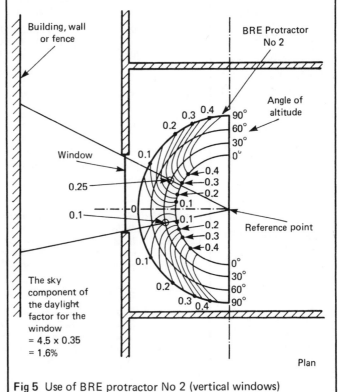

Fig 5 Use of BRE protractor No 2 (vertical windows)

Notes

The external reflected component of the daylight factor for a uniform sky may be taken as approximately one-tenth of the equivalent sky component. Using the diagrams shown in Daylighting 2 the value may be found as follows

(a) Readings on protractor No 1 are 4% and 0.5%;

(b) Equivalent sky component = 4 − 0.5 = 3.5%;

(c) Average angle of altitude = 15°;

(d) Readings on auxiliary protractor No. 2 are 0.27 and 0.13.

(e) Correction factor = 0.27 + 0.13 = 0.4 ∴ Equivalent uniform sky component = 3.5 x 0.4 − 1.3%.

(f) Externally reflected component = one-tenth of 1.3% = 0.13%.

2   The last stage in calculating the daylight factor is to determine the light reaching the reference point after reflection from the surfaces inside the room i.e. internal reflected component. Light surfaces have high reflection factors; dark surfaces have low reflection factors (see Table 2).

Table 1 Reflection factors

Reflection factors (%)		Reflection factors (%)	
White	75–88	Golden yellow	62
Light stone	53	Orange	36
Middle stone	37	Eau-de-nil	48
Light buff	60	Sky blue	47
Middle buff	43	Turquoise	27
Light grey	44	Light brown	30
Dark grey	26	Middle brown	20
Pale cream	73	Salmon pink	42

Table 2 Minimum internally reflected component of the daylight factor (%)

Ratio of window area to floor area	Window area as a percentage of floor area	Floor reflection factor (%)											
		10				20				40			
		Wall reflection factor (per cent)											
		20	40	60	80	20	40	60	80	20	40	60	80
		%	%	%	%	%	%	%	%	%	%	%	%
1:50	2			0.1	0.2			0.1	0.2		0.1	0.2	0.2
1:20	5	0.1	0.1	0.2	0.4	0.1	0.2	0.3	0.5	0.1	0.2	0.4	0.6
1:14	7	0.1	0.2	0.3	0.5	0.1	0.2	0.4	0.6	0.2	0.3	0.6	0.8
0:10	10	0.1	0.2	0.4	0.7	0.2	0.3	0.6	0.9	0.3	0.5	0.8	1.2
1:6.7	15	0.2	0.4	0.6	1.0	0.2	0.5	0.8	1.3	0.4	0.7	1.1	1.7
1:5	20	0.2	0.5	0.8	1.4	0.3	0.6	1.1	1.7	0.5	0.9	1.5	2.3
1:4	25	0.3	0.6	1.0	1.7	0.4	0.8	1.3	2.0	0.6	1.1	1.8	2.8
1:3.3	30	0.3	0.7	1.2	2.0	0.5	0.9	1.5	2.4	0.8	1.3	2.1	3.3
1:2.9	35	0.4	0.8	1.4	2.3	0.5	1.0	1.8	2.8	0.9	1.5	2.4	3.8
1:2.5	40	0.5	0.9	1.6	2.6	0.6	1.2	2.0	3.1	1.0	1.7	2.7	4.2
1:2.2	45	0.5	1.0	1.8	2.9	0.7	1.3	2.2	3.4	1.2	1.9	3.0	4.6
1:2	50	0.6	1.1	1.9	3.1	0.8	1.4	2.3	3.7	1.3	2.1	3.2	4.9

Note: The ceiling reflection factor is assumed to be 70%

Example. Find the minimum internally reflected component of the daylight factor for a room measuring 10 m x 8 m x 2.5 m high, having a window in one wall with an area of 20 m². The floor has an average reflection factor of 20% and the walls and ceiling average reflection factors of 60% and 70% respectively

Window area as a percentage of floor area $= \frac{20}{80} \times \frac{100}{1} = 25\%$

Referring to Table 2 the minimum internally reflected component = 1.1%. Allowing a maintenance factor of 0.9 for dirt on the windows the value will be modified to 1.1 x 0.9 = 0.99%. For the example given in daylighting 2 and 3 the daylight factor will be the addition of the three components = 0.35 + 0.13 + 0.99 = 1.47% approx.

25 LIFTS AND ESCALATORS

PLANNING OF LIFTS

Notes

1   In order to function efficiently, multi-storey buildings must be provided with a correctly designed lift installation. The The planning of lift installations (as with all services) should commence early in the design project.

2   Lifts should be sited in the central area of the building and account must be taken of the positions of entrances and staircases. When a building has a number of passenger lifts they should be grouped together. In some large buildings it may be necessary to have a main group of lifts near the main entrance and a single lift at the other end of the building.

3   The lift lobby must be wide enough to allow traffic to move past the lift without causing congestion.

4   For tall buildings express lifts may be required.

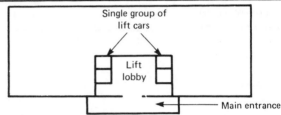

Fig 1 Building with a single group of lifts

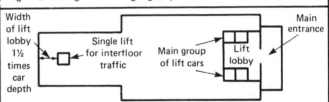

Fig 2 Building with a main group of lifts and also a single lift serving interfloor traffic

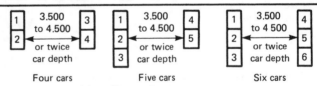

Fig 3 Groups of four five or six cars

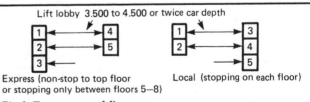

Express (non-stop to top floor or stopping only between floors 5–8)   Local (stopping on each floor)

Fig 4 Two groups of five cars

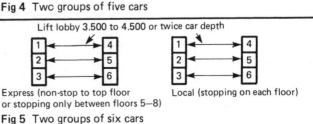

Express (non-stop to top floor or stopping only between floors 5–8)   Local (stopping on each floor)

Fig 5 Two groups of six cars

ROPING SYSTEMS FOR ELECTRIC LIFTS — 1

Notes

1 High tensile steel ropes are used having a factor of safety of ten. The ropes travel over grooved driving or traction sheaves and pulleys. The counter weight reduces the load on the electric motor.

2 The single-wrap 1:1 roped is the most economical and efficient roping system. It is used for a small car.

3 The single-wrap 1:1 with diverter pulley, is required for a large car and diverts the counterweight away from the car. To prevent rope slip, the sheave and pulley may be double wrapped.

4 The single wrap 2:1 roped is again used for a large car. The system doubles the load carrying capacity of the machine but requires more rope and also reduces the car speed by 50%.

5 Double wrapped roping improves the traction between the Counterweight driving sheave and the steel ropes.

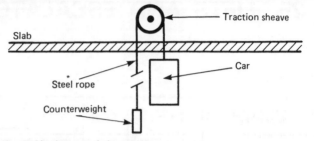

Fig 6 Single wrap 1:1 roped

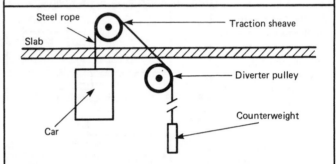

Fig 7 Single wrap 1:1 roped with diverter pulley

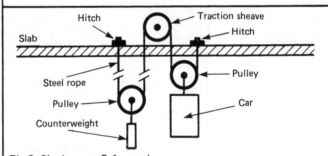

Fig 8 Single wrap 2:1 roped

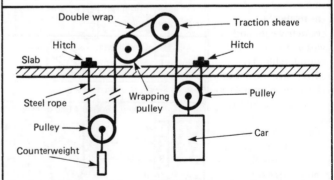

Fig 9 Double wrap 2:1 roped (for high speed and medium to heavy duty loads)

ROPING SYSTEMS FOR ELECTRIC LIFTS — 2

Notes

The single wrap 3:1 roping system is used for heavy goods lifts where it is necessary to reduce the force acting upon the bearings and the counterweights electrical power required for the motor. The system trebles the load carrying capacity of the machine but requires a good deal of rope and also reduces the car speed by about 66.6%.

2   In the drum drive, one set of ropes is wound clockwise around the drum and the other set anti-clockwise. The disadvantage of the drum drive is that, as the height increases, the drum becomes too unwieldy and is rarely used for rises of over 30 m.

3   In buildings above 10 storeys the weight of rope transferred from the car to the counterweight and vice versa, is considerable. To offset this unbalanced load and to reduce the risk of the car bouncing a compensating rope is required. The double wrap reduces the risk of rope slip.

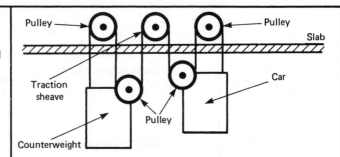

Fig 10 Single wrap 3:1 roping

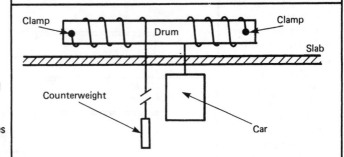

Fig 11 Drum drive

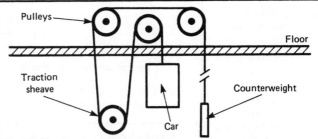

Fig 12 Single wrap 1:1 roped with machine room below roof level. The length of rope is increased which limits the travel and speed of car

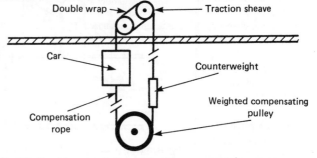

Fig 13 Double wrap 1:1 roped with compensating rope

SINGLE AUTOMATIC LIFT CONTROL

Notes

The single automatic push button is the simplest of the control systems. The lift car can be called and used by only one person or a group of persons at a time. When the lift car is called to a floor the signal lights engraved 'in use' are illuminated on every floor. The car will not then respond to any subsequent landing calls, nor will these landing calls be recorded and stored. The car is under complete control of the occupant until he or she has reached the required floor and has left the car.

2   The 'in use' lights are now switched off and the car available to answer the next landing call registered.

3   Although the control is simple and cheap when compared with other systems, an intending passenger must wait until the car is stationary and empty before it will respond to a call.

4   The system is most suitable for light traffic conditions in buildings such as nursing homes, small hospitals and low rise flats.

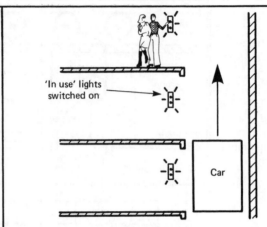

'In use' lights switched on

Car

Car unoccupied and responding to the first landing call

Fig 14 Lift car called to a floor. 'In use' lights switched on

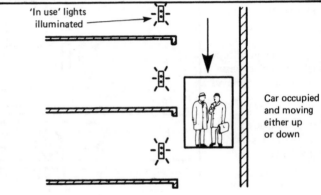

'In use' lights illuminated

Car occupied and moving either up or down

Fig 15 Lift car in control of occupant and cannot be called by other passengers

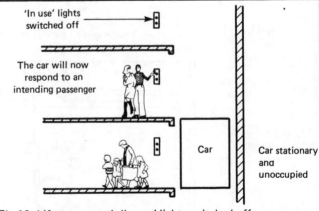

'In use' lights switched off

The car will now respond to an intending passenger

Car

Car stationary and unoccupied

Fig 16 Lift car vacated. 'In use' lights switched off. Lift can now be called by other passengers

170

DOWN COLLECTIVE LIFT CONTROL

Notes

1   The down collective control system stores calls made by passengers in the cars and on the landings. These calls are answered in a logical order to ensure the minimum car movement.

2   If the lift car is moving upwards, it will respond to calls made inside the car in floor sequence. When the car reaches the highest registered call it will automatically reverse and move down answering all the landing calls in floor sequence during its descent.

3   If the car also stores up and down landing calls and answers them in logical order the system is known as the full collective control system. When the car is moving upwards this system will respond to up car calls and up landing calls. It will also respond to both down car calls and down landing calls.

4   The down collective system is used for light duty buildings where traffic is mainly upwards from the main floor and downwards from upper floors e.g. flats and small hotels.

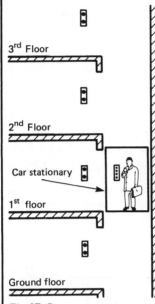

Fig 17 Passengers enter the car and press buttons to travel upwards

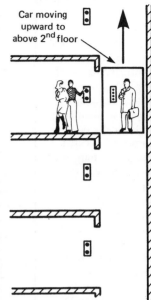

Fig 18 While travelling upwards all the landing calls are by-passed

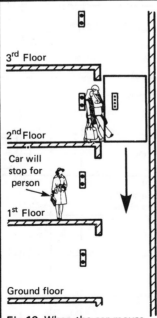

Fig 19 When the car moves down all landing calls are collected floor by floor

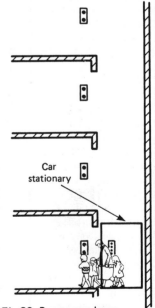

Fig 20 Passengers leave the car.

CONTROLS FOR TWO, THREE AND EXCEEDING THREE CARS

Notes

1   Two cars may be co-ordinated by a central system so that the efficiency of the lifts is increased. Each car operates individually on a full or down collective control system. When the cars are at rest, however, one car will be stationed on the main floor and the other car, which is free, will be stationed on a mid-point floor or any other convenient floor. The free car will answer landing calls on any floor, except the main floor. If, however, the free car remains in use for a given period and unable to answer a landing call, the other car, if free, will answer the call.

2   A similar system may also be used for three cars with two cars stationary on the main floor and one free car at mid point or top floor.

3   In Fig 23 each car operates on full collective control and each car will respond to calls within its own zone. A computer dispatches the cars in each zone.

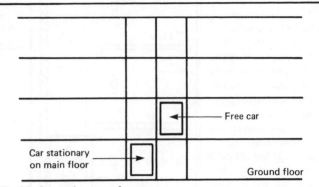

Fig 21 Control system for two cars

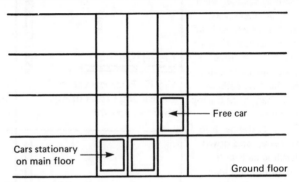

Fig 22 Control for three cars

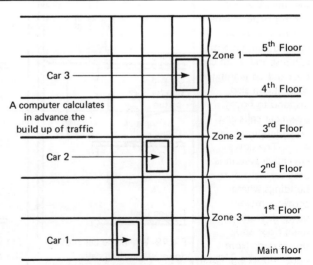

Fig 23 Supervisory control for three or more cars

LIFT MACHINE ROOM AND EQUIPMENT

Notes

1   Wherever possible the machine room should be sited above the lift shaft as this position reduces the lengths of the ropes and increases the efficiency. The room should be ventilated but the vent opening must not be over the machine. The machinery should be well secured to a concrete slab and to reduce the transmission of sound compressed cork slabs should be inserted between the machinery and the concrete.

An overhead beam should be fixed over the machinery for positioning or removing the equipment for repair.

Enough floor space should be provided for the inspection and repair of the equipment.

To reduce the risk of condensation the room should be well insulated. The room air temperature should not rise above 40°C nor fall below 10°C. The walls, ceiling and floor should be smooth and painted to reduce the formation of dust.

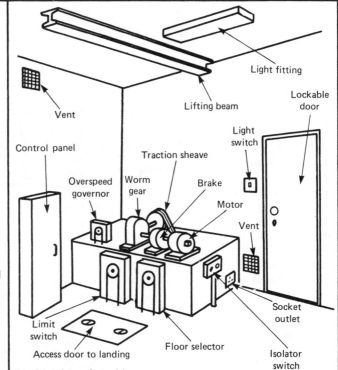

Fig 24 View of machine room

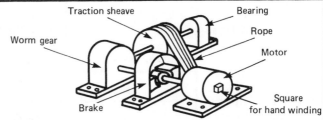

Fig 25 View of geared traction machine (for car speeds up to 0.8 m/s)

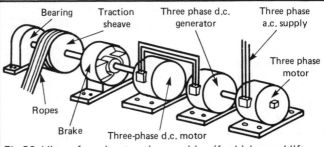

Fig 26 View of gearless traction machine (for high speed lifts, 1.75 m/s and over)

DETAILS OF AN ELECTRIC LIFT INSTALLATION

Notes

1   The size of the lift shaft will depend upon the size of the car; space will be required for the counterweight, guides and landing door. The shaft must extend below the lowest level served in the form of a pit. The pit permits overtravel of the car and buffers are required at the base of the pit for both the car and counterweight. The pit must be made watertight and drainage provided.

2   The shaft and pit must be plumb and the internal surfaces finished smooth and painted to minimise collection of dust. A smoke vent having an unobstructed area of 0.1 m² must be inserted at the top of the shaft. The shaft should be of either reinforced concrete 130 mm thick (min) or brick 230 mm thick (min).

3   No pipes, ventilating ducts or cables (other than the lift cables) must be fixed inside the shaft.

4   A clearance at the top of the shaft is required for overtravel of the car. The counterweight may be at the back or at the side of the car.

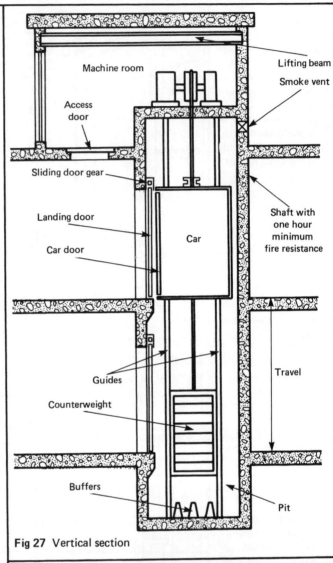

Fig 27 Vertical section

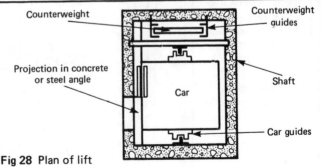

Fig 28 Plan of lift

174

OIL-HYDRAULIC LIFTING ARRANGEMENTS

Notes

1   The direct acting method is the simplest and most effective method but it requires a borehole for the ram; the ram may be in one piece or telescopic. The width of the shaft is kept to the minimum which saves on the construction costs.

2   In the side-acting method, the ram is connected to the side of the car. For heavy goods lifts two rams may be used, one ram at each side of the car. A borehole is not required but owing to cantilever design, when a single ram is used, there are limitations on the car size and load carried.

3   In the direct side-acting method the car is cantilevered and suspended by a steel rope. The cantilever design limits the size of the car and load carried, the car speed however may be increased.

4   In the indirect side-acting method the car is centrally suspended by a steel rope and the hydraulic system is inverted.

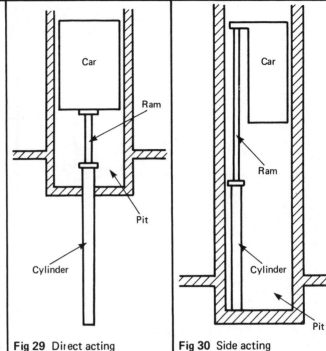

Fig 29 Direct acting

Fig 30 Side acting

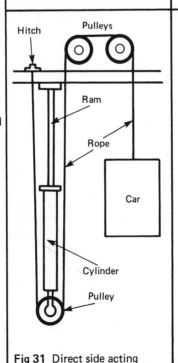

Fig 31 Direct side acting

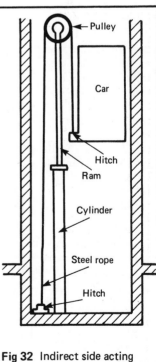

Fig 32 Indirect side acting

DETAILS OF OIL-HYDRAULIC LIFT INSTALLATION

Notes

1   An oil-hydraulic lift is the modern version of the older type hydraulic lift which was supplied with water from a water main. The water main was pressurised from a central pumping station and several buildings were supplied for the lift installation. In the oil-hydraulic system, oil is forced under pressure into a cylinder thus raising a ram and the lift car.

2   Each lift has its own pumping unit and controller. These units are usually sited at the lowest level served but they may also be sited at some distance from the lift.

3   The lift is ideal where moderate speeds are required and the travel distance is not great. The car speed ranges from 0.12 m/s and 1 m/s and the maximum travel is usually 21 m. The lift is particularly suitable for goods lifts and for hospitals and old persons' homes.

4   Most oil-hydraulic lifts carry the load directly to the ground and therefore the shaft does not carry the lift loads. The construction of the shaft is therefore cheaper. Acceleration and travel is very smooth.

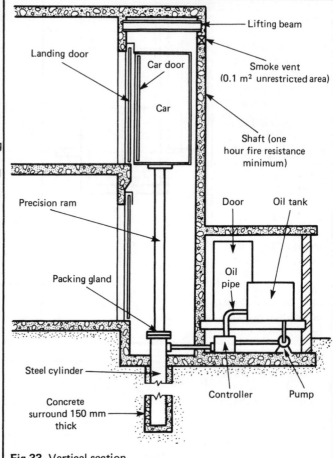

Fig 33 Vertical section

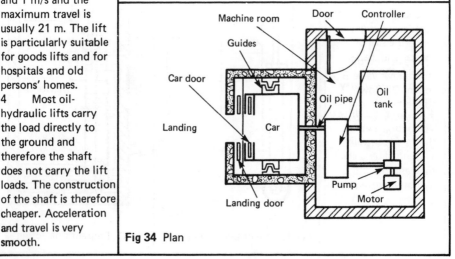

Fig 34 Plan

OIL-HYDRAULIC LIFT PUMPING UNIT AND PACKING GLAND

Notes

1   When the car is required to move upwards the oil pressure must be gradually increased. The up solenoid valve is energised by an electric current and opens to allow oil to enter above piston D. Since the area of piston D is greater than valve C the oil pressure closes the valve and allows high pressure oil to flow to the cylinder and lift the ram and the car.

2   When the car is required to move downwards, the oil pressure must be gradually decreased. The lowering solenoid valve is energised by an electric current and opens allowing oil to flow back to the tank through the by-pass. Since the area of piston A is greater than valve B the reduction of oil pressure behind the piston allows valve B to open. Oil flows into the tank and the car moves downwards.

3   A special packing gland is required between the cylinder and ram.

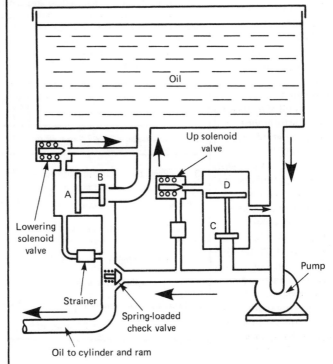

Fig 35 Oil tank, pump and controls

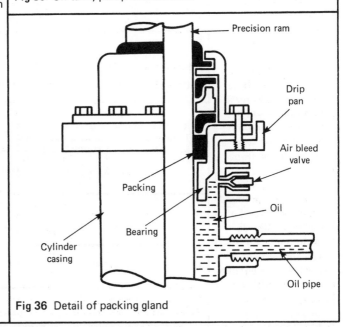

Fig 36 Detail of packing gland

NUMBER OF LIFTS REQUIRED

Notes

1 Lift performance depends on:
(a) acceleration;
(b) retardation,
(c) car speed;
(d) speed of door operation; and
(e) stability of speed and performance with variations of car load.

2 The assessment of population density may be found by allowing between one person per 9.5 m² of floor area to 11.25 m² of floor area. For unified starting and finishing times 17% of the population per five minutes may be used. For staggered starting and finishing times 12% of the population may be used.

3 The number of lifts will have an effect on the quality of service. Four 16-person lifts provide the same capacity as three 24-person lifts but the waiting time will be about twice as long with the three-car group.

4 The quality of service may be found from the interval of the group. 25–35 seconds interval is excellent, 35–45 seconds is acceptable for offices 60 seconds for hotels and 90 seconds for flats.

Example. A twenty-storey office block having unified starting and stopping times is to have a floor area above the ground floor of 8000 m² and room heights of 2.4 m. A group of four lifts, each car having a capacity of 20 persons and a car speed of 2.5 m/s are specified. The clear door width is to be 1.1 m and the doors are to open at a speed of 0.4 m/s. Estimate the interval and quality of service that is to be provided.

1 Peak demand for a 5 minute period $= \dfrac{8000 \text{ m}^2 \times 17\%}{11 \text{ m}^2/\text{person} \times 100}$

     "   "   " " "   "   " $= 124$ persons

2 Car travel $= 20 \times 2.4 \text{ m} = 48 \text{ m}$

3 Probable number of stops $= S - S\left(\dfrac{S-1}{S}\right)^n$
(where S = maximum number of stops)

$\therefore$ Probable number of stops $= 20 - 20\left(\dfrac{20-1}{20}\right)^{16}$

   "   "   "   " $= 11$
(where n = number of passengers usually approximately 80% of capacity)

4 Upward journey time $= S_1\left(\dfrac{L}{S_1 V} + 2V\right)$

where S_1 = probable number of stops L = travel V = speed

$\therefore$ Upward journey time $= 11\left(\dfrac{48}{11 \times 2.5}\right) + 2 \times 2.5$

   "   "   " $= 24.25$ seconds

5 Downward journey time $= \left(\dfrac{L}{V} + 2V\right)$

   "   "   " $= \dfrac{48}{2.5} + 2 \times 2.5$

   "   "   " $= 24.2$ seconds

6 Door operating time $= 2\,(S_1 + 1)\,\dfrac{W}{Vd}$

where W = width of door opening; Vd = opening speed

$\therefore$ Door operating time $= 2\,(11 + 1)\,\dfrac{1.1}{0.4} = 66$ seconds

7 The average time taken for each person to get into and out of a lift car may be taken as two seconds
$\therefore$ Transfer time $= 2n = 2 \times 16 = 32$ seconds

8 Round trip time $= 24.25 + 24.2 + 66 + 32 = 146.45$ seconds

9 Capacity of group $= \dfrac{5 \text{ mins} \times 60 \times 4 \times 20 \times 0.8}{146.45}$

   "   "   " $= 131$ persons per 5 minutes

10 Interval for the group $= \dfrac{146.45}{4} = 36.6$ seconds

The capacity of the group of lifts and the interval for the group are satisfactory. (*Note*: Cars less than 12 capacity are not satisfactory)

PATERNOSTERS

Notes

1   A paternoster consists of a series of open fronted two-person cars suspended from hoisting chains. The chains run over sprocket wheels at the top and bottom of the lift shaft. The lift is continuously moving and provides for both up and down movement of passengers in one shaft. The passengers enter or leave the car while it is moving and therefore the waiting time is reduced to the minimum. The passengers however, have to be fairly agile and the lift is therefore suitable for use in factories, universities and colleges. It is not suitable for the infirm or elderly.

2   When a car reaches the limit of its travel in one direction, it moves across to the other set of hoisting chains and engages with car guides to travel in the other direction. The speed of the lift car must not exceed 0.4 m/s.

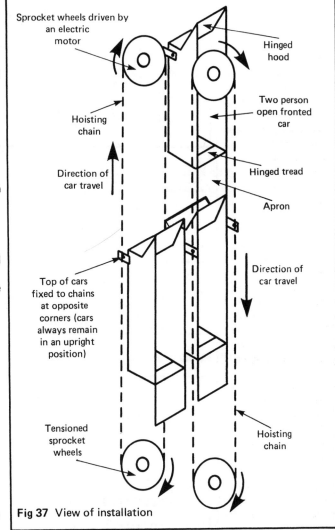

Fig 37 View of installation

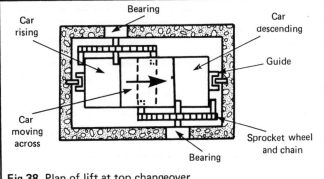

Fig 38 Plan of lift at top changeover

179

ESCALATORS

Notes

1 Escalators are used for moving people from one floor level to another. A pair of escalators, one up and one down, are used and these can handle between them up to 12 000 persons per hour. The maximum carrying capacity depends upon the width of the steps and their speed. The step widths vary between 600 mm and 1.2 m and the step speed varies between 0.45 m/s and 0.85 m/s.

2 Sophisticated control gear, such as is required for a lift, is unnecessary since the motor runs continuously and the load varies gradually. This helps to reduce the cost of maintenance and possible breakdown.

3 The escalator may be supplied in one or two units which reduces site work.

4 To prevent the spread of fire through the floor opening a water sprinkler may be installed to automatically provide a curtain of water over the well. Alternatively, a fire-proof shutter operated by a smoke detector or fusible links, may be used.

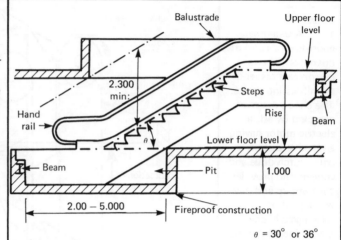

θ = 30° or 36°

Fig 39 Elevation

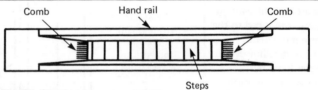

Fig 40 Plan

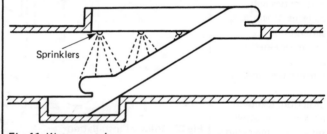

Fig 41 Water curtain

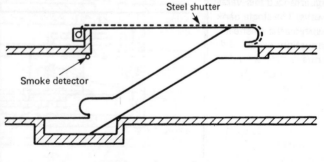

Fig 42 Fireproof sliding shutter

LIFT DOORS AND ESCALATOR ARRANGEMENTS

Notes

1   The landing doors are mechanically interlocked and are operated by the car doors. The doors are operated by an electric motor having a speed reduction unit, clutch drive and connecting mechanism. The type of entrance and doors form a vital part of the lift installation. The average lift car will spend more time at a floor during passenger transfer time than it will during travel. For normal passenger service, either side opening or two-speed side opening doors are used. The most efficient in terms of passenger handling is the two-speed centre opening. The clear opening may be greater and usable clear space becomes more quickly available to the passengers.

2   Escalators may be arranged in several ways. The single bank with traffic in one direction avoids interruption of traffic but takes up more floor space than other arrangements. A criss-cross arrangement is used for traffic moving in both directions.

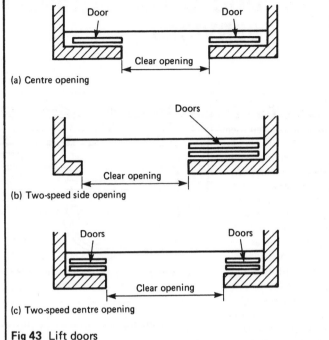

(a) Centre opening

(b) Two-speed side opening

(c) Two-speed centre opening

Fig 43 Lift doors

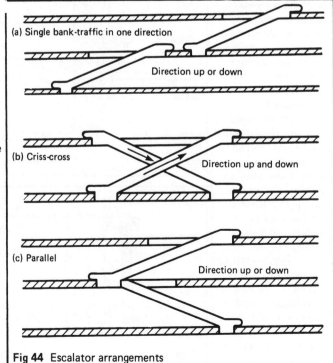

(a) Single bank-traffic in one direction

Direction up or down

(b) Criss-cross

Direction up and down

(c) Parallel

Direction up or down

Fig 44 Escalator arrangements

26 DUCTS FOR ENGINEERING SERVICES

DUCTS FOR ENGINEERING SERVICES

Notes

1   Before laying ducts for the entry of services into the building, it is essential to ascertain the positions of the pipes or cables belonging to the public utilities and to provide entry for the shortest practicable routes for the building pipes or cables.

2   For flexible pipes or cables, a pipe duct containing a bend may be used. For rigid pipes or large cables, a straight pipe duct to a pit will be required.

It is essential that the pipe duct is sealed at the inlet end or otherwise water may pass to the underside of the site concrete.

3   For housing horizontal pipes or cables, a skirting or floor duct may be used. The services may be supported by 'build in' brackets or saddle clips. Vertical pipes or cables may be housed in either a surface type duct or a chase.

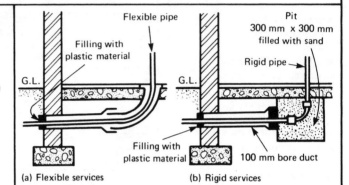

(a) Flexible services   (b) Rigid services

Fig 1 Ducts for entry of services into the building

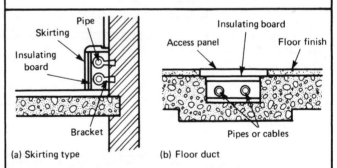

(a) Skirting type   (b) Floor duct

Fig 2 Horizontal ducts for small pipes or cables

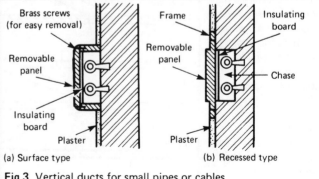

(a) Surface type   (b) Recessed type

Fig 3 Vertical ducts for small pipes or cables

MEDIUM AND LARGE VERTICAL DUCTS

Notes

1 The purpose of a service duct is to conceal the services and to facilitate inspection, repair and alterations of the services. A service duct also helps to reduce noise and protects the services from damage.

2 In designing a service duct, the transmission of noise, possible build-up of heat, spread of fire and accessibility to the services must be considered.

3 Vertical ducts usually extend the full height of the building, the number of ducts required will depend upon the number of services and the positions of the equipment served.

4 The duct must have a fire resistance of at least one hour. Where the duct passes through the floor, the space around the duct should be fire stopped to the full depth of the floor.

5 Segregation of services must be carried out in accordance with CP. 413 : 1973.

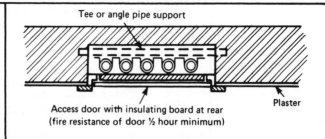

Tee or angle pipe support

Access door with insulating board at rear
(fire resistance of door ½ hour minimum)

Plaster

Fig 4 Recessed for medium-sized pipes and cables

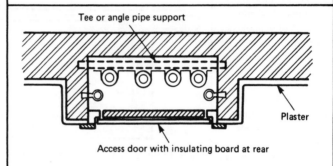

Tee or angle pipe support

Access door with insulating board at rear

Plaster

Fig 5 Partially recessed for medium-sized pipes and cables

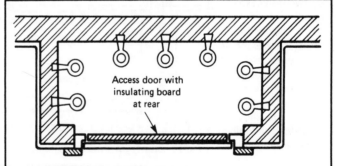

Access door with insulating board at rear

Fig 6 Built-out for large pipes

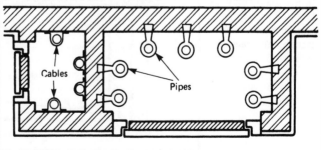

Cables

Pipes

Fig 7 Built-out for large pipes and cables

MEDIUM AND LARGE HORIZONTAL DUCTS

Notes

1   Access openings to floor trenches and crawl-ways are usually provided at intervals, or continuous covers may be fitted. A crawl-way should be wide enough to allow a clear working space of at least 700 mm.

2   Trench covers (for continuous access) may be of timber, stone, reinforced concrete, metal, or in the form of metal trays filled to match the floor finish.

The covers should be light enough to be lifted by one man, or by two men at the most.

3   Sockets for lifting handles should be incorporated in the covers. In external situations, the cover slabs (usually of stone or concrete) should be bedded and joined together with lime mortar.

4   If timber covers are used to match a timber floor, they may be fixed with brass cups and screws. A trench must have an internal depth of less than 1 m and a crawl-way an internal depth of 1 m (min).

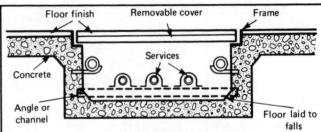

Fig 8 Floor trench with removable cover

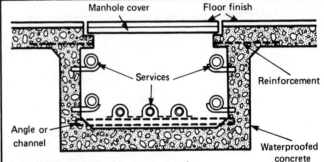

Fig 9 Floor trench with access opening

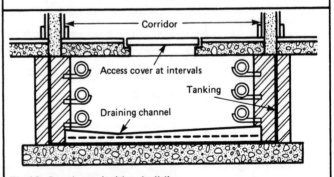

Fig 10 Crawl-way inside a building

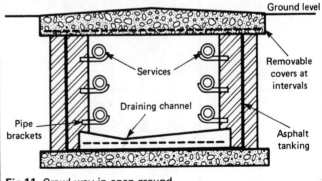

Fig 11 Crawl-way in open ground

SUBWAYS OR WALKWAYS

Notes

1   The main access to a subway will normally be gained from the boiler room control room or basement, but additional access should be provided at convenient points, especially at junctions and changes of directions. Removable manhole covers strong enough to carry any loads or traffic should be provided at these points.

Ducts which house services from a boiler or control rooms must be provided with a self-closing fire door at the entry.

2   It is desirable that ducts are ventilated (preferably to the open air). Ducts below ground should be provided with a shallow channel to convey water from whatever source, to a sump or drain. The connection to a drain should be through a sealed gully. This will prevent the entry of sewer gas, vermin or back-flow. The position of the gully must be clearly marked on the side of the duct.

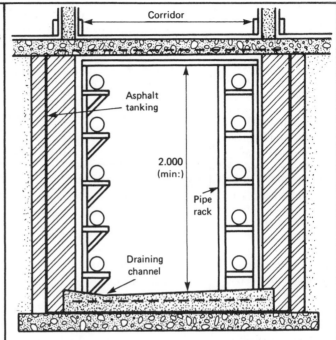

Fig 12 Subway inside a building

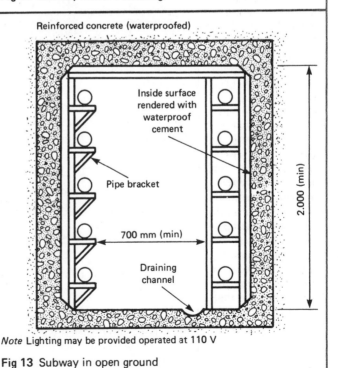

Note Lighting may be provided operated at 110 V

Fig 13 Subway in open ground

PENETRATION OF STRUCTURE BY PIPES

Notes

1   The risk of fire spread associated with pipework inside a building can be divided into four areas:
(a) addition of fuel to the total fire load;
(b) production of toxic gases and smoke;
(c) risk of fire spread along the pipework;
(d) reduction of fire resistance of building elements penetrated by services.

2   The Building Regulations 1991, B 3, give the requirements of penetration of structure by pipes (see Fig 15).

Pipes of lead or aluminium or an alloy of these metals, asbestos cement or unplasticed polyvinyl chloride pipes complying with BS4519: 1969 may have a maximum internal diameter of 100 mm. Other types of pipe materials may have a maximum internal diameter of 40 mm.

3   If a metal pipe is passed through the structure, a plastic pipe connected to it may have the same specification as the metal pipe, providing that the plastic pipe is at least 1 m from the structure.

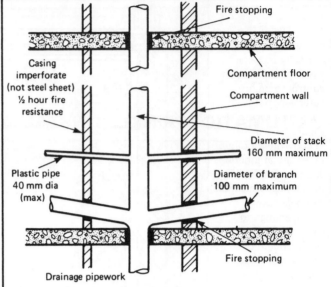

Fig 14 Pipes inside a protected shaft

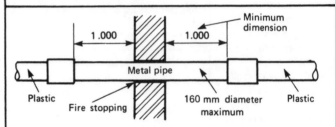

Fig 15 160 mm diameter uPVC pipe through separating wall, protected structure, compartment wall or floor (unless wholly enclosed in a protected shaft or cavity barrier)

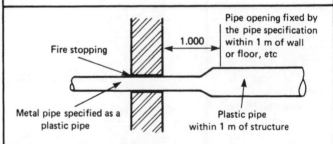

Fig 16 Pipe of different materials and diameters through separating wall, protected structure, compartment wall or floor (unless wholly enclosed in a protected shaft or cavity barrier)

186

27 AUTOMATIC SPRINKLER SYSTEMS

AUTOMATIC SPRINKLER INSTALLATIONS — 1

Notes

1 The wet pipe sprinkler system is used in heated buildings where there is no risk of the water freezing. The pipes are filled with water under pressure at all times and the sprinkler heads are usually fixed below the range pipes. If mains water is used it should be fed from both ends. If there is a burst main on one side, the main stop valve and branch pipe can be closed and the sprinkler system supplied from the other branch pipe.

2 When a sprinkler head opens by heat from a fire, the water flows through the annular groove in the alarm valve seating through to the pipe connected to the alarm gong and turbine. A jet of water strikes the turbine blades and the turbine revolves causing the alarm gong to operate. The main stop valve must be locked in an open position.

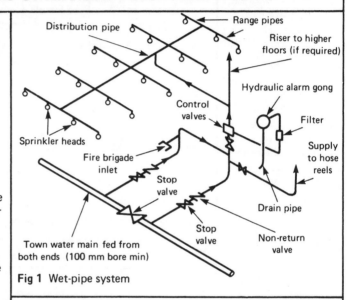

Fig 1 Wet-pipe system

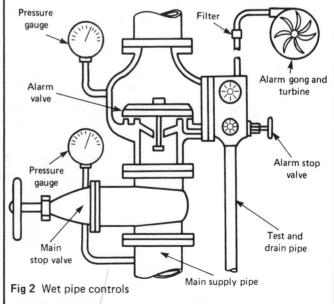

Fig 2 Wet pipe controls

AUTOMATIC SPRINKLER INSTALLATIONS — 2

Notes

1   Either dry or alternate wet-and-dry sprinkler systems are usually installed in buildings that are without a heating system.

The dry system pipework, above the differential valve, is always filled with compressed air. When a fire occurs the compressed air escapes, and the differential valve opens allowing water to flow to the open sprinkler or sprinklers.

2   The alternate wet-and-dry system works as a wet system during summer and as a dry system during winter. It is usual practice to charge the systems with air at a gauge pressure of about 138 kPa. A small compressor is used to restore any loss of air pressure.

The compressor will start automatically when the air pressure in the pipework falls but will have little effect on the flow of water to the sprinklers.

3   A pump switches on automatically when the sprinklers open and will improve the flow of water to the sprinklers. The heads must be fitted above the range pipes so that the system may be drained.

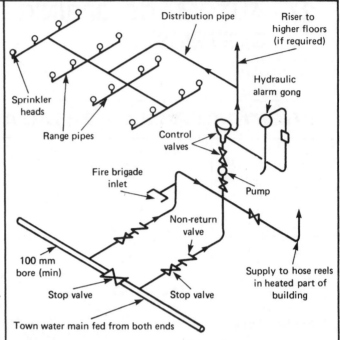

Fig 3 Dry pipe or alternate wet-and-dry pipe system

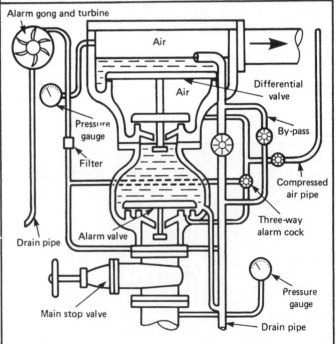

Fig 4 Dry pipe or alternate wet-and-dry pipe controls

TYPES OF SPRINKLER HEAD

Notes

1   The bulb-type head contains a bulb made of special glass which is used to hold the water valve in place. The bulb contains a special coloured liquid which, when heated to a certain temperature, expands sufficiently to shatter the glass and the valve opens. Water flows on to the deflector and sprays over the fire. The temperature at which the sprinkler opens ranges from 57° C Orange, 68° C Red, 79° C Yellow, 93° C Green, 141° C Blue, 182° C Mauve and 227 to 288° C Black.

2   The fusible soldered strut type head contains metal struts soldered together which are used to hold the water valve in place. At certain temperatures heat from a fire melts the solder and the struts fall away allowing the head to open and water to spray on to the fire.

3   The Duraspeed soldered type head contains a heat collector on to which a cap is soldered. When heat from a fire melts the solder the cap falls away thus displacing a strut which allows the head to open.

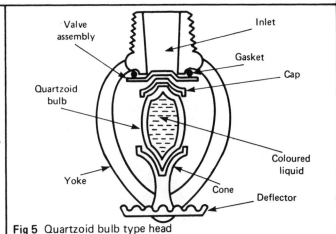

Fig 5 Quartzoid bulb type head

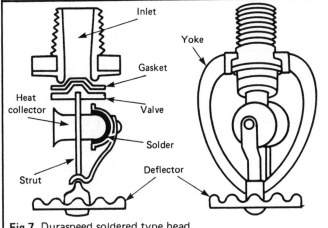

Fig 6 Fusible soldered strut type head

Fig 7 Duraspeed soldered type head

189

DELUGE AND MULTIPLE CONTROL SPRINKLERS

Notes

1   The deluge system is used for special fire hazards such as plastic foam-making machines, firework factories, aircraft hangars, etc where intensive fires with a very fast rate of fire propagation are expected.

When a fire occurs the quartzoid bulbs shatter and the compressed air in the pipeline is released. This causes a rapid fall in air pressure on a diaphragm inside the automatic deluge valve to which both compressed air and water are connected. The movement of the diaphragm causes the deluge valve to open and water to discharge on to the fire, through the projector.

2   In the multiple control sprinkler system, a heat sensitive sealed valve controls the flow of water to a small group of sprayers which operate simultaneously.

When a fire occurs the quartzoid bulb shatters and allows the valve stem to fall thus opening the valve. Water then flows through the small group of sprayers which cover the protected area.

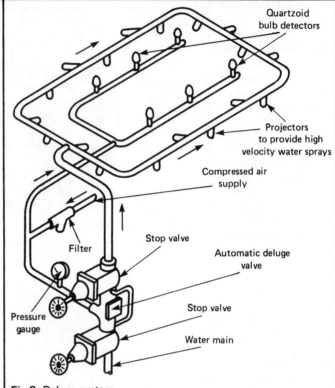

Fig 8 Deluge system

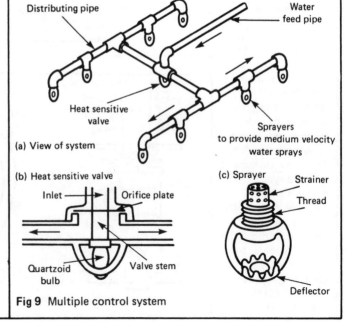

(a) View of system

(b) Heat sensitive valve

(c) Sprayer

Fig 9 Multiple control system

METHODS OF DISTRIBUTION TO SPRINKLERS

Notes

1   The method by which the distribution pipework to the sprinklers is arranged will depend upon the building layout and the position of the riser pipe. In order to provide an even distribution to the sprinkler heads it is always preferable to have a centre feed pipe. In practice, however, this is not always possible and an end feed pipe may have to be used.

The maximum spacing of the sprinkler heads (S) on the range pipes depends upon the fire hazard classification of the building. The standard spacings are: (a) extra light hazard 4.6 m; (b) ordinary hazard 4.0 m; (c) extra high hazard 3.7 m.

2   The maximum floor areas covered by one sprinkler head are: (a) extra light hazard 21 m²; (b) ordinary hazard 12 m²; (c) extra high hazard 9 m². Extra light hazards include offices, hotels hospitals. Ordinary hazards include restaurants and engineering factories. Extra high hazards are carpets, foamed plastics and wood.

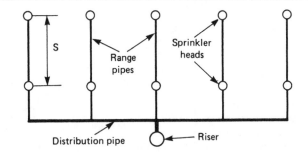

Fig 10 Two-end side with centre feed pipe

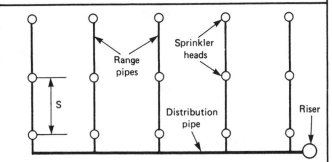

Fig 11 Three-end side with end feed pipe

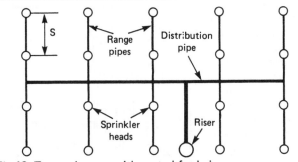

Fig 12 Two-end centre with central feed pipe

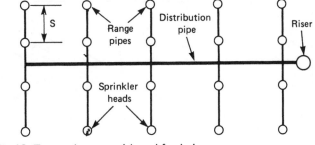

Fig 13 Two-end centre with end feed pipe

WATER SUPPLY FOR SPRINKLER SYSTEMS

Notes

1 The efficiency of a sprinkler system depends largely upon the type of water supply used. An elevated private reservoir must have a minimum volume of water of between 9 m³ and 875 m³.

2 A suction tank supplied from the main with automatic pumping may be used providing that the tank has a minimum volume of water of between 2.5 m³ and 585 m³. A higher standard of cover may be provided by combining the suction tank with a pressure tank gravity tank or elevated private reservoir. The pressure tank must have a minimum volume of water of between 7 m³ and 23 m³. A pressure switch or a flow switch automatically cuts in the pump when the sprinklers open.

3 A gravity tank is usually sited on a tower high enough to provide sufficient head of water above the sprinklers.

4 If water is taken from a river or canal the strainers must be fitted well below the lowest possible water level.

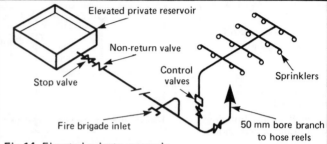

Fig 14 Elevated private reservoir

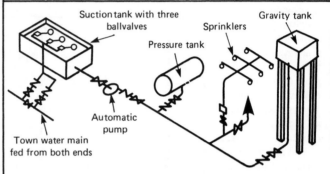

Fig 15 Town main suction tank automatic pump with pressure tank or gravity tank (if required)

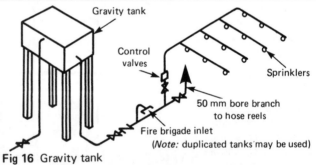

Fig 16 Gravity tank
(containing between 9 m³ and 875 m³ of water)

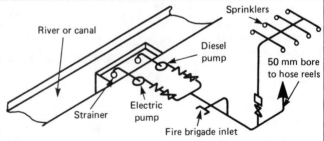

Fig 17 Automatic pumps drawing from river or canal

DRENCHERS

Notes

1   A drencher fire control system provides a discharge of water over roofs, walls and windows to prevent fire spreading from nearby buildings. Automatic drenchers are similar to quartzoid bulb sprinklers and operate individually on the same principle. Non-automatic drenchers have open nozzles and a manually controlled valve is opened to bring them into operation.

A drain valve is required at the lowest point and the pipes must fall to this point. The stop valve must be placed in a prominent position where access to it will not be impeded by fire.

2   As in the case of sprinklers, two drenchers can be supplied by a 25 mm bore pipe. A 50 mm bore pipe can supply ten drenchers, a 76 mm bore pipe thirty six drenchers and a 150 mm bore pipe over one hundred drenchers.

3   In theatres, drenchers may be fitted above the proscenium arch at the stage side for the protection of the safety curtain.

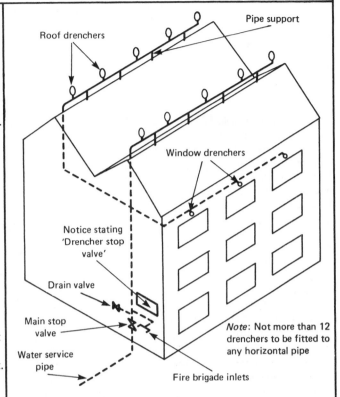

Fig 18 Typical drencher installation

Note: Not more than 12 drenchers to be fitted to any horizontal pipe

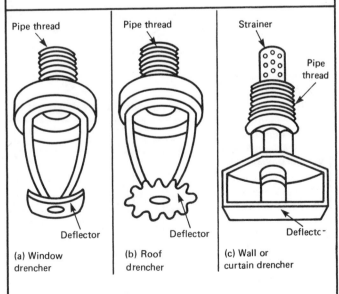

(a) Window drencher

(b) Roof drencher

(c) Wall or curtain drencher

Fig 19 Types of drencher

193

28 HOSE REELS AND FIRE RISERS

HOSE REEL INSTALLATIONS

1   A hose reel is a fire-fighting appliance for use by the occupants of the building. Hose reels should be sited in positions where they can be used without exposing the users to danger from fire e.g. the staircase landing.

2   The hose should be able to discharge 0.4 litre per second at a distance of 6 m from the end of the nozzle. Three reels should be capable of being used simultaneously.

A pressure of 200 kPa is required at the highest reel and if the town water main will not provide this pumping will be required.

If a suction tank is required it should hold a minimum volume of water of 1.6 m³. A 50 mm bore pipe is required for buildings up to 15 m in height and a 64 mm pipe is required for buildings above 15 m.

3   Hose reels may (continued on page 195)

(continued on page 195)

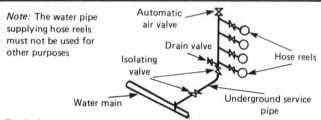

Note: The water pipe supplying hose reels must not be used for other purposes

Fig 1 Supply to hose reels direct from main

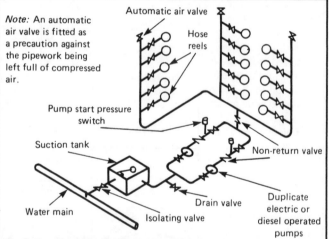

Note: An automatic air valve is fitted as a precaution against the pipework being left full of compressed air.

Fig 2 Supply to hose reels indirect from main

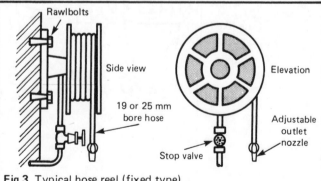

Fig 3 Typical hose reel (fixed type)

DRY RISER

1   Risers should be disposed so that no part of the floor is more than 61 m from a landing valve. This distance should be measured along a route suitable for the hose line including any distance up and down the stair way.

2   In buildings up to 45 m in height with one 64 mm landing valve on each floor, a 100 mm bore riser is required. In buildings between 45 m and 60 m in height, with one or two 64 mm landing valves on each floor, a 150 mm bore riser is required. For buildings above 60 m in height, a wet riser must be installed above 60 m. Two 64 mm bore inlets are required for a 100 m bore riser and four 64 mm bore inlets are required for a 150 mm bore riser. The riser must be electrically earthed.

(continued from page 194) be either of the fixed or swinging types and are installed about 1 m above the floor. Reinforced non-kink rubber hose is used in lengths from 18 m to 46 m. A 25 mm pipe should supply each reel.

Note: A dry riser is installed either in unheated buildings or where the water main will not provide sufficient pressure at the highest landing valve. The riser is, in effect, an extension of the fire brigade hose. A hard standing for the Fire Engine is required at the base of the riser. One landing valve is required for every 930 m² of floor area

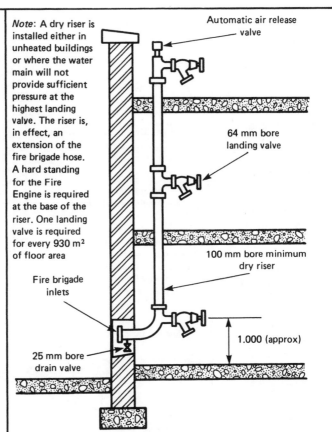

Automatic air release valve

64 mm bore landing valve

100 mm bore minimum dry riser

1.000 (approx)

25 mm bore drain valve

Fire brigade inlets

Fig 4 Typical arrangement of a dry riser

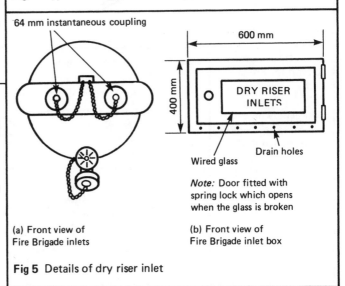

64 mm instantaneous coupling

600 mm

400 mm

DRY RISER INLETS

Drain holes

Wired glass

Note: Door fitted with spring lock which opens when the glass is broken

(a) Front view of Fire Brigade inlets

(b) Front view of Fire Brigade inlet box

Fig 5 Details of dry riser inlet

WET RISER

Notes

1   A wet riser is constantly filled with water and the supply should be capable of maintaining a running pressure of 410 kPa at the highest landing valve and give a flow rate of 23 litres per second. The maximum running pressure permitted with one outlet open should be 510 kPa.

2   In order to maintain the required water pressure it is usually necessary to install pumping equipment. Direct pumping from the main is not permitted and a suction tank with a minimum actual volume of water of 45.5 m² must be installed.

3   One landing valve (64 mm bore) should be provided for every 930 m² of floor area.

4   It is necessary to limit the pressure on the canvas hose to 660 kPa. This is achieved by means of a pressure relief valve incorporated in the outlet of the landing valve. The discharge from the pressure relief valve should be carried by a 100 mm bore drain pipe.

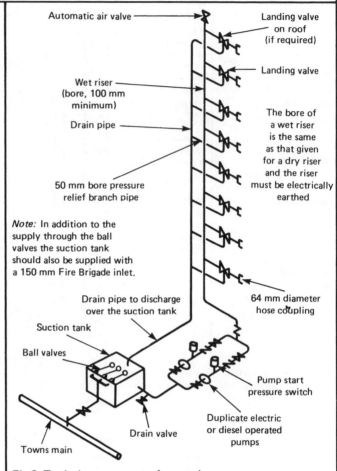

Automatic air valve

Landing valve on roof (if required)

Landing valve

Wet riser (bore, 100 mm minimum)

Drain pipe

The bore of a wet riser is the same as that given for a dry riser and the riser must be electrically earthed

50 mm bore pressure relief branch pipe

Note: In addition to the supply through the ball valves the suction tank should also be supplied with a 150 mm Fire Brigade inlet.

Drain pipe to discharge over the suction tank

Suction tank

Ball valves

64 mm diameter hose coupling

Pump start pressure switch

Duplicate electric or diesel operated pumps

Drain valve

Towns main

Fig 6 Typical arrangement of a wet riser

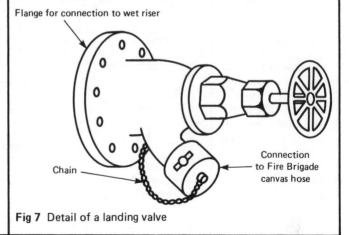

Flange for connection to wet riser

Chain

Connection to Fire Brigade canvas hose

Fig 7 Detail of a landing valve

29 PORTABLE AND FIXED FIRE EXTINGUISHERS

PORTABLE FIRE EXTINGUISHERS — 1

Notes

1   A portable fire extinguisher must contain the type of fire extinguishing agent suitable for the fire it may be required to extinguish. Fires are classified in four groups:

Class A. Fires in solid fuels, e.g. wood, paper, cloth, etc.

Class B. Fires in inflammable liquids, e.g. petrol, oil, paints, fats, etd.

Class C. Fires in inflammable gases, e.g. methane, propane, acetylene, etc.

Class D. Fires in inflammable metals, e.g. zinc, aluminium, uranium etc.

Class E. Fires caused by an electrical fault.

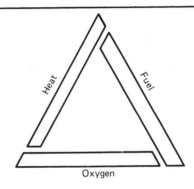

Fig 1   Three elements required for a fire. The removal of one element will extinguish the fire

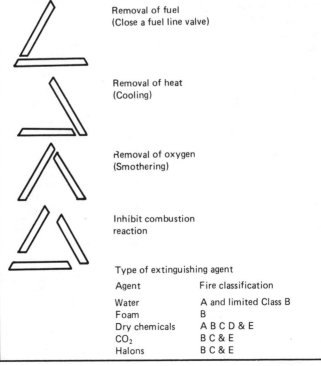

Removal of fuel
(Close a fuel line valve)

Removal of heat
(Cooling)

Removal of oxygen
(Smothering)

Inhibit combustion reaction

Type of extinguishing agent

Agent	Fire classification
Water	A and limited Class B
Foam	B
Dry chemicals	A B C D & E
CO_2	B C & E
Halons	B C & E

PORTABLE FIRE EXTINGUISHERS — 2

Notes

2   Water is often used for extinguishing Class A fires and water-type extinguishers are commonly installed in offices, schools, hotels, etc. In the striking type soda-acid portable extinguisher sulphuric acid inside a glass bottle is released when a knob is struck. The sulphuric acid mixes with water plus carbonate of soda and a chemical reaction takes place which produces carbon dioxide gas. The cylinder is thus pressurised and liquid is forced out of the nozzle. The inversion type operates on the same principle but sulphuric acid is released by inversion.

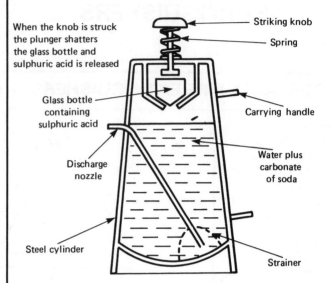

When the knob is struck the plunger shatters the glass bottle and sulphuric acid is released

Striking knob

Spring

Glass bottle containing sulphuric acid

Carrying handle

Discharge nozzle

Water plus carbonate of soda

Steel cylinder

Strainer

Fig 2(a) Striking type soda-acid water portable fire extinguisher

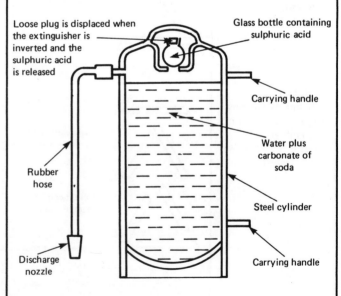

Loose plug is displaced when the extinguisher is inverted and the sulphuric acid is released

Glass bottle containing sulphuric acid

Carrying handle

Water plus carbonate of soda

Rubber hose

Steel cylinder

Discharge nozzle

Carrying handle

Fig 2(b) Inversion type soda-acid water portable fire extinguisher

PORTABLE FIRE EXTINGUISHERS — 3

Notes

1 In order for a fire to occur three things must be present, (a) heat; (b) oxygen (c) fuel: If one of these is removed, the fire will be extinguished.

Water removes heat but cannot be used for some types of fires. Oxygen can be removed by smothering the fire with a fire blanket, foam or carbon dioxide. Foam may be used for gas or liquid fires.

In the chemical foam type of extinguisher, foam is formed by chemical reaction between sodium bicarbonate and aluminium sulphate dissolved in water and in the presence of a foaming agent. When the extinguisher is inverted the chemicals are mixed thus forming foam under pressure which is forced out of the nozzle.

2 In the carbon dioxide extinguisher, liquid CO_2 is pressurised inside a cylinder and when the disc is pierced the liquid is forced out through the nozzle as a gas.

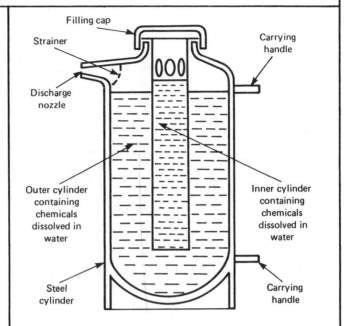

Fig 3 Chemical foam portable fire extinguisher (inversion type)

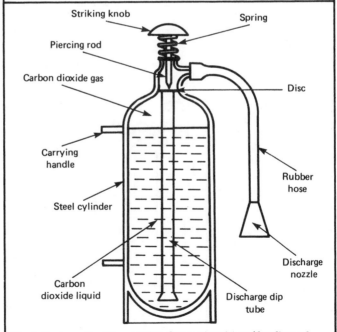

Fig 4 Carbon dioxide portable fire extinguisher (for fires of liquids and gases and electrical fires)

FIXED FOAM INSTALLATIONS

Notes

1   A pump-operated mechanical foam installation consists of a foam concentrate tank sited outside the area to be protected. The tank has a water supply pipe inlet and a foam pipe outlet. A venturi is fitted in the pipeline to draw the foam out of the tank. When the water pump is switched on, the venturi effect causes a reduction in pressure at the foam pipe connection and this results in a mixture of foam concentrate and water being discharged through the outlet pipe. The foam solution is then discharged over the protected area.

2   A pre-mix foam installation consists of a storage tank containing foam solution which, when being discharged, is under a pressure of about 1000 kPa.

When a fire occurs in the protected area, a fusible link is broken which allows a weight to fall and open a valve on the carbon dioxide cylinder. Foam solution is forced out of the tank to discharge over an oil tank, etc.

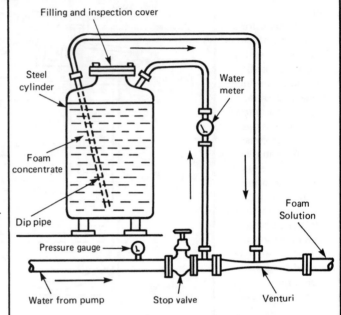

Fig 5 Pump operated mechanical foam installation

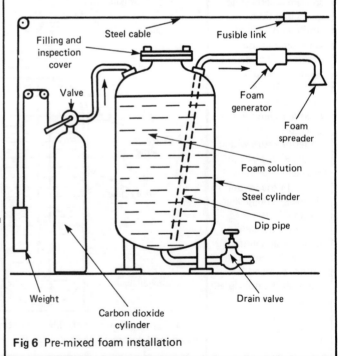

Fig 6 Pre-mixed foam installation

FOAM INSTALLATIONS

Notes

1 A foam installation is used for the application of foam from remote points on to fire risks involving flammable liquids. This type of installation is used mainly in connection with oil-fired boilers and oil storage tanks for heating systems. A foam box is built into the wall of the building at a point approved by the fire authority and where it is easily accessable. The box is usually placed about 600 mm above ground level and should be clear of any openings through which heat, smoke or flame can pass. A glass front can be broken and the lock released from the inside. Two 64 mm diameter inlets may be used.

2 A 64 mm diameter pipe is normally used and all pipework must slope slightly towards the spreader. Vertical drop pipes may be of any length but vertical riser pipes must not be used. The length of pipe (A) must not be less than 300 mm and both legs (H) must be equal. Foam spreaders (S) are 64 mm diameter with twin elbows.

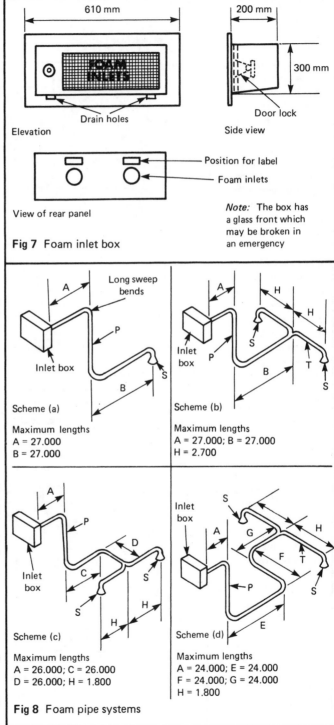

Elevation

Drain holes

Side view

Door lock

View of rear panel

Position for label

Foam inlets

Fig 7 Foam inlet box

Note: The box has a glass front which may be broken in an emergency

Scheme (a)

Long sweep bends

Inlet box

Maximum lengths
A = 27.000
B = 27.000

Scheme (b)

Inlet box

Maximum lengths
A = 27.000; B = 27.000
H = 2.700

Scheme (c)

Inlet box

Maximum lengths
A = 26.000; C = 26.000
D = 26.000; H = 1.800

Scheme (d)

Inlet box

Maximum lengths
A = 24.000; E = 24.000
F = 24.000; G = 24.000
H = 1.800

Fig 8 Foam pipe systems

FIXED HALON AND CARBON DIOXIDE INSTALLATIONS

Notes

1   Halon 1301 is an electrically non-conductive gas and personnel can safely remain in the protected area during discharge.

The system is very suitable for areas where a high density of equipment is present such as tape libraries and computer rooms. The gas is stored in spherical steel cylinders which can be fixed in a ceiling void or against a wall. Smoke actuates the detector which immediately opens valves on the extinguisher and total flooding of the protected area occurs.

2   As an alternative to Halon 1301, carbon dioxide gas may be used. This gas is dry, non-conductive and is heavier than air so it flows around obstacles. It can be used for the protection of transformers, computer rooms, textile machinery, etc. Integrated high and low pressure gas systems may be used. The smoke detectors are used to open valves on the cylinders.

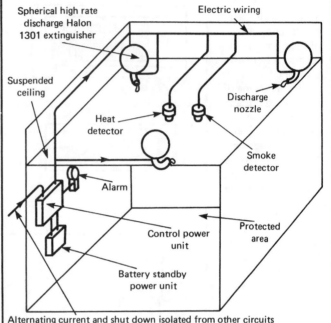

Fig 9 Halon 1301 installation

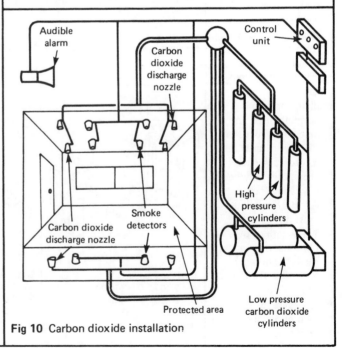

Fig 10 Carbon dioxide installation

30 FIRE DETECTORS

SMOKE DETECTORS

1   In Fig 1 a positively charged plate attracts negative ions and a negatively charged plate attracts positive ions. An ion is an atom or a group of atoms which have lost or gained one or more electrons and thus carries a negative or a positive charge. The movement of the ions between the plates reduces the resistance of the air so that a small electric current is produced. During a fire, smoke enters the detector and particles of smoke become attached to the ions and slow their movement. This reduces the flow of current which actuates the alarm circuit.

2   In Fig 2(a) the light beam does not fall on a photo-electric cell. When smoke enters the detector some of the light beam is deflected upwards onto the photo-electric cell. An electric current is thus produced which actuates the alarm.

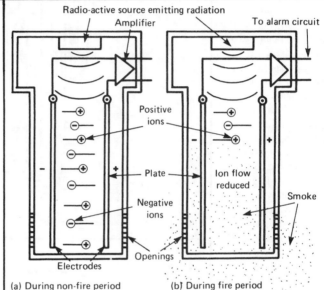

(a) During non-fire period   (b) During fire period

Fig 1 Ionisation smoke detector

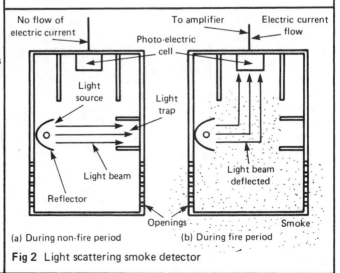

(a) During non-fire period   (b) During fire period

Fig 2 Light scattering smoke detector

HEAT DETECTORS

Notes

1   Heat detectors are used where smoking is permitted and where a smoke detector would be actuated prematurely. The detector is designed to detect a fire in its more advanced stage so it is not as efficient as a smoke detector.

The fusible alloy detector contains a thin walled case fitted with heat collecting fins at its lower end. A conductor passes through the centre. The case is line with a fusible alloy, which acts as a second conductor. Heat from a fire melts the fusible alloy at a predetermined temperature and causes it to make contact with the central conductor. This completes an electrical circuit to sound the alarm.

2   In the bi-metal coil detector heat from a fire passes through the cover to the bi-metal coils and the lower coil receives greater heat initially than the upper coil. The lower coil quickly makes contact with the upper coil to complete the electrical circuit and sound the alarm.

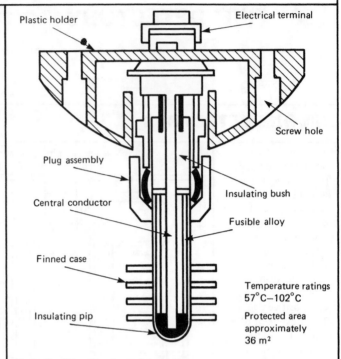

Plastic holder

Electrical terminal

Screw hole

Plug assembly

Central conductor

Insulating bush

Fusible alloy

Finned case

Insulating pip

Temperature ratings
57°C–102°C

Protected area
approximately
36 m²

Fig 3 Fusible alloy heat detector

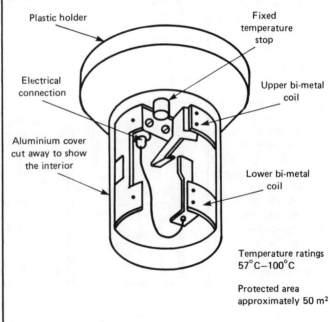

Plastic holder

Fixed temperature stop

Electrical connection

Upper bi-metal coil

Aluminium cover cut away to show the interior

Lower bi-metal coil

Temperature ratings
57°C–100°C

Protected area
approximately 50 m²

Fig 4 Bi-metal coil heat detector

LIGHT OBSCURING AND LASER BEAM DETECTORS

Notes

1 In the light-obscuring fire detector system a beam of light is projected across the protected area close to the ceiling. The light falls on to a photo-electric cell and a small current of electricity is produced which is amplified before being connected to the alarm circuit.

If a fire occurs smoke rises and passes through the light beam which is obscured and interrupts the light reaching the photo-electric cell. The flow of electric current from the cell is stopped and the alarm signal is actuated.

2 A laser beam is a band of light which can be visible or infra-red, it does not fan out or diffuse as it travels. The beam will operate over a distance of up to 100 m. If a fire occurs smoke or heat rises and the pulsating beam is deflected away from the photo-electric cell. The flow of electric current is stopped an and the alarm signal is actuated.

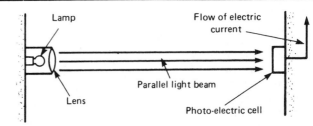

(a) Detector during non-fire period

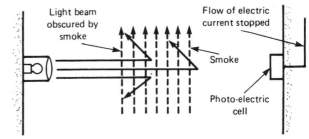

(b) Detector during fire period

Note: The light beam will operate over a distance up to 15 m.

Fig 5 Light obscuring detector

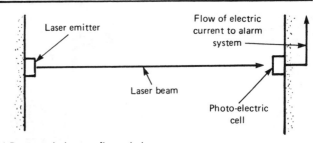

(a) Detector during non-fire period

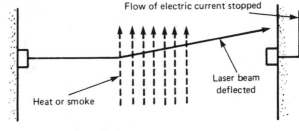

(b) Detector during fire period

Fig 6 Laser beam detector

205

RADIATION FIRE DETECTORS

Notes

1 Beside producing hot gases, a fire also releases radiant energy in the form of visible light, infra-red and ultra-violet radiation. Radiant energy travels in waves from the fire and radiation detectors are designed to respond to this energy.

The infra-red detector uses a photo-electric cell and the lens and filter will allow only infra-red radiation to be used. Flames have a distinctive flicker, normally in the range of 4 Hz to 15 Hz and the filter/amplifier is used to both amplify the current and also to filter out signals not in this range. To reduce false alarms a timing device operates the alarm a few seconds after an outbreak of fire.

2 The ultra-violet detector uses a gas-filled bulb, which when struck by ultra-violet radiation, ionises the gas in the bulb and produces an electric current. When the current flow is greater than the set point of the amplifier, the alarm circuit closes immediately to operate the alarm bell.

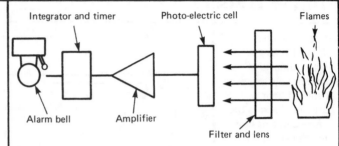

Fig 7 Components of an infra-red detector

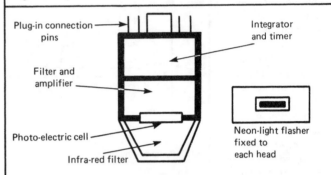

Fig 8 Infra-red detector for small areas

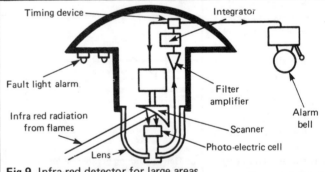

Fig 9 Infra-red detector for large areas

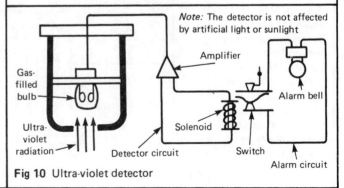

Fig 10 Ultra-violet detector

FIRE DETECTOR ELECTRICAL CIRCUITS

Notes

1   In an open circuit the call points or detectors are connected to open switches and there is therefore no current flow through the circuit when it is on standby. The operation of a call point or a detector closes the switches and actuates the alarm. Because there is no current flow in the circuit when on standby, it does not consume as much electrical power as a closed circuit. A broken circuit, however, will prevent some detectors from operating.

2   In a closed circuit the call points or detectors may be regarded as closed switches thus allowing current to flow in the detector circuit. When a call point or a detector is operated, a solenoid-operated switch is prevented from being energised and a spring closes the switch. This allows current to flow through the alarm circuit and actuate the alarm.

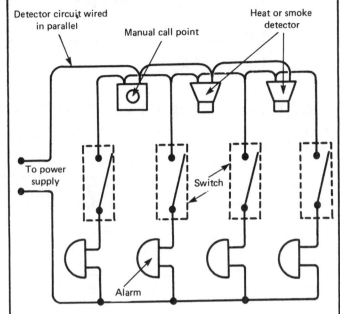

Fig 11 Diagram of 'open' alarm circuit

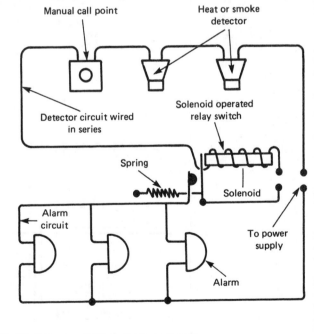

Fig 12 Diagram of 'closed' alarm circuit

31 FIRE PREVENTION IN DUCTWORK AND SMOKE CONTROL

FIRE PREVENTION IN VENTILATING SYSTEMS

Notes

1   In the event of fire passing into a vertical enclosure the smoke and hot gases may fill the space and the temperature at the top may be sufficient to set fire to any combustible materials and seriously distort the services. The enclosure can be ventilated to the outside by an opening of not less than 0.05 m². If the enclosure should be divided at every third floor by means of non-combustible material for the full thickness of the floor. The enclosure should be gas-tight and there must be no access to the enclosure from a staircase.

2   If the ventilation ductwork is not within a fire-resisting enclosure, fire dampers must be fitted at points where the ductwork passes from one compartment to another.

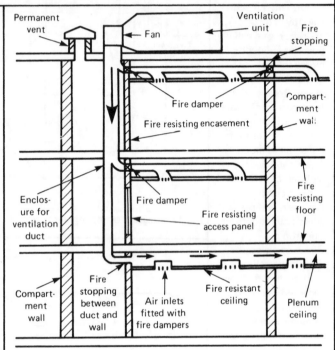

Fig 1 Installation of ventilating ductwork

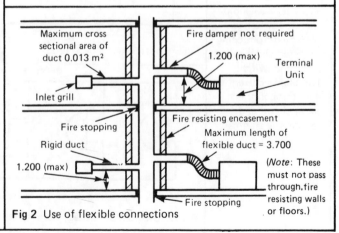

Fig 2 Use of flexible connections

FIRE DAMPERS USED IN VENTILATING SYSTEMS

Notes

1   Fire dampers are required in ducts to prevent the spread of fire in a building. Mechanical dampers may be operated by either a fusible link or an electro-magnet device. This is normally operated by a smoke detector which can be arranged to operate one or all the dampers. If the mechanical dampers are operated by a fusible link, the link breaks at a predetermined temperature usually 70°C and a steel plate or shutter seals the duct.

2   In the intumescent-coated honeycomb damper the paint expands when one hundred times its original volume and forms a mass thus preventing the passage of smoke or fire through the duct. The paint is not affected by fluff or oil spray but it should be kept free from greasy dirt, condensation and damp which will interfere with its effectiveness. The damper is normally used where the air velocity is low.

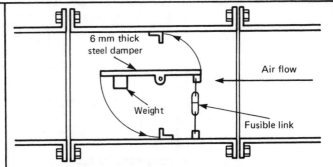

Fig 3 Swinging mechanical type

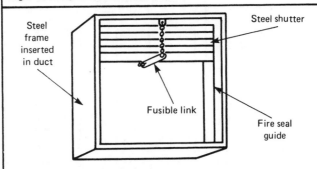

Fig 4 Sliding mechanical type

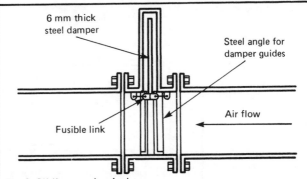

Fig 5 Shutter mechanical type

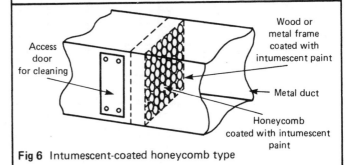

Fig 6 Intumescent-coated honeycomb type

PRESSURISATION OF ESCAPE ROUTES

Notes

1   In multi-storey buildings staircases and lobbies may be pressurised to clear smoke and thus provide an easy escape route. The air pressurisation level is usually between 25 and 50 Pa depending upon the building height and degree of exposure.

   Three methods are used:

(a) pressurisation plant normally off but is automatically switched on in the event of a fire;

(b) pressurisation plant runs on full duty during all hours of occupancy;

(c) pressurisation plant runs continuously at reduced capacity during all hours of occupancy but is automatically brought to full duty during the event of a fire.

2   It is important to provide openings so that smoke can pass from the escape route to the outside air. The system permits siting the lobby and staircase in a central position and thus provides greater freedom in planning. It also prevents the entry of rain and draughts from staircase openings.

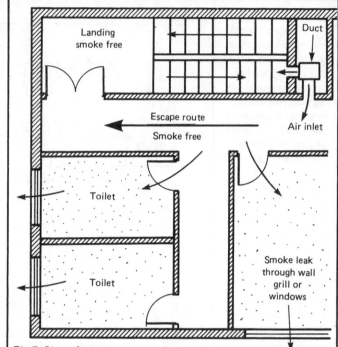

Fig 7 Plan of escape route and rooms

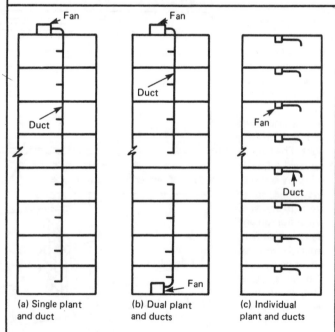

(a) Single plant and duct

(b) Dual plant and ducts

(c) Individual plant and ducts

Fig 8 Methods of installing ductwork

210

FIRE VENTILATION

Notes

1   Automatic fire ventilation is designed to remove heat, smoke and toxic gases from single storey buildings. In large single-storey factories the additional volume of air entering the building by fire venting is not sufficient to significantly increase the rate of fire. Portions of the roof are divided into sections by use of fireproof screens and fire vents are fitted at the highest part of the roof of each section.

2   If a fire occurs, heat rises in the roof section above the fire. At a predetermined temperature, usually 70°C, a fusible link breaks and opens the ventilator above the fire. Heat and smoke pass through the ventilator and smoke logging of the building is thus prevented. Firemen can see the fire and can extinguish it without the use of breathing apparatus. The heat removed prevents the risk of an explosion, flash over and distortion of the steel truss.

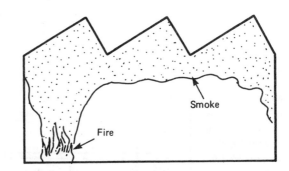

Fig 9 Fire in unvented building showing unrestricted spread of smoke

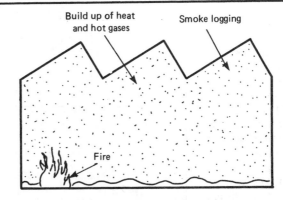

Fig 10 Fire in unvented building showing ultimate smoke logging

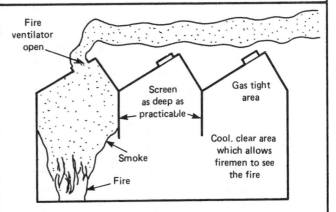

Fig 11 Fire in vented building showing restricted spread of smoke. The fire ventilator may also be used for normal ventilation.

SMOKE CONTROL IN COVERED SHOPPING CENTRES

Notes

1   Most enclosed shopping centres have a mall with a parade of shops. The mall forms the normal circulation area and is therefore the obvious escape route in the event of fire. A fire in a shop or in the mall can cause a rapid spread of smoke and hot gases. Therefore, it is necessary that some form of smoke control must be adopted. If the centre is sprinklered the water may cool the smoke and hot gases so reducing their buoyancy and possible fogging at floor level.

If sprinklers are used they should operate at a higher temperature than normal to reduce the number of heads and thus lower the risk of loss of buoyancy by cooling.

2   Smoke can be controlled by:
(a) providing smoke reservoirs into which the smoke can flow through before being extracted by either mechanical or natural means;
(b) allowing cooler air to enter the centre at low level to replace the smoke flowing out of the centre.

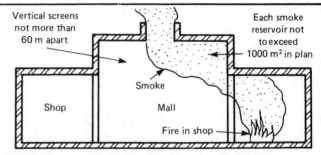

Fig 12 Smoke reservoir by adopting a greater ceiling height in the mall than in the shops

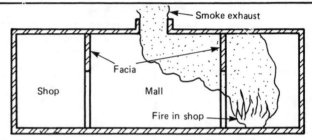

Fig 13 Smoke reservoir formed by facias above open fronted shops

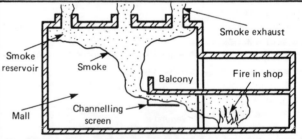

Fig 14 Two-storey mall showing behaviour of smoke through channelling screens

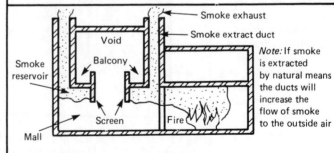

Fig 15 Use of smoke extract ducts through roof of mall

212

APPENDIX 1
CONDENSING GAS BOILERS

This is the most efficient boiler. It has a larger heat exchanger for transferring heat to the water in the heating system.

The boiler extracts heat from the flue gases by allowing them to condense in the boiler. Unlike an ordinary boiler the efficiency remains high even when working at a low output.

The bench efficiency is about 95 per cent whereas an ordinary boiler has a bench efficiency of about 74 per cent.

A drainage pipe is needed for condensate disposal.

Savings range from 15p to 20p in the pound.

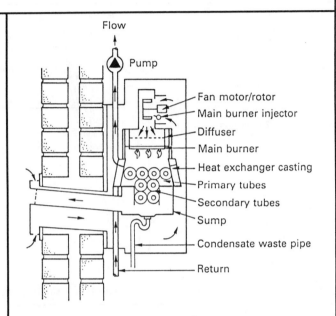

Flow

Pump

Fan motor/rotor

Main burner injector

Diffuser

Main burner

Heat exchanger casting

Primary tubes

Secondary tubes

Sump

Condensate waste pipe

Return

Fig 1 Balanced flue condensing boiler

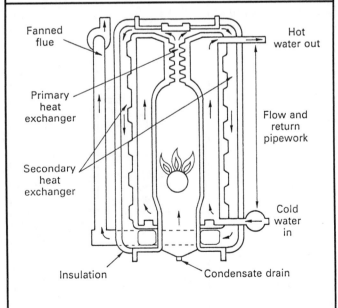

Fanned flue

Hot water out

Primary heat exchanger

Flow and return pipework

Secondary heat exchanger

Cold water in

Insulation

Condensate drain

Fig 2 Conventional flue condensing boiler

APPENDIX 2 ECONOMY 7 ELECTRIC WATER HEATING

Because most industry and commerce close down at night-time and there is spare capacity of electricity, this off-peak electricity is therefore cheaper.

Economy 7 operates from 12 midnight to 7 a.m. It is about half the cost of ordinary day-time tariff.

Because the aim of Economy 7 is to rely on heating the water during the off-peak hours a larger hot water cylinder is required and 144 or 210 litres is recommended.

The hot water cylinder must be well insulated. Clock and thermostatic controls are required.

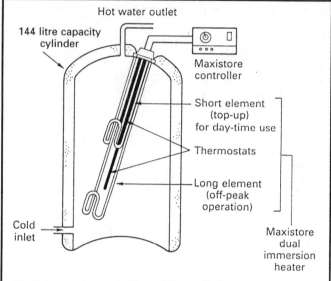

Fig 1 Immersion heater for existing cylinder

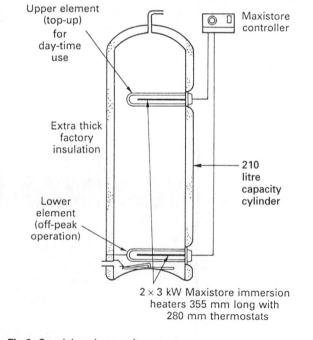

Fig 2 Special package unit

APPENDIX 3
COMPACT FLUORESCENT LAMPS

Compact fluorescent lamps are miniature fluorescent lamps which last over 8000 hours rather than 1000 hours for an ordinary lamp, and use only 20-25 per cent of the energy. They can make a positive contribution to protecting the environment.

The comfort type gives a gentle diffused light and can be used for long burning purposes.

The prismatic types are more robust and can be used in workshops and commercial purposes.

Electronic types are the most efficient and use only 20 per cent of the energy of ordinary light lamps. The lamps are not suitable with a dimmer switch.

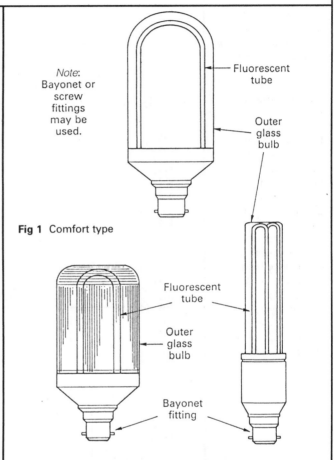

Note: Bayonet or screw fittings may be used.

Fluorescent tube

Outer glass bulb

Fig 1 Comfort type

Fluorescent tube

Outer glass bulb

Bayonet fitting

Fig 2 Prismatic type

Fig 3 Electronic type

Energy Saving Chart			
Energy saver	Ordinary lightbulb	Energy saving	Over 8000 hours save up to
25W	100W	80%	£47.70
18W	75W	73%	£36.25
11W	60W	80%	£31.16
9W	40W	72%	£19.72
Domestic energy cost 7.95p/kWh			

APPENDIX 4 THE COMBI SYSTEM

Now that VAT has been imposed on fuel, a combi boiler will help to reduce its cost.

The system does not require cisterns in the roof space, there is no hot water cylinder or connecting pipework and therefore less heat losses.

The combi gas boiler only heats the quantity of water required and the water is at mains pressure thus providing a good shower. The boiler may be sited in the airing cupboard giving more space in the kitchen.

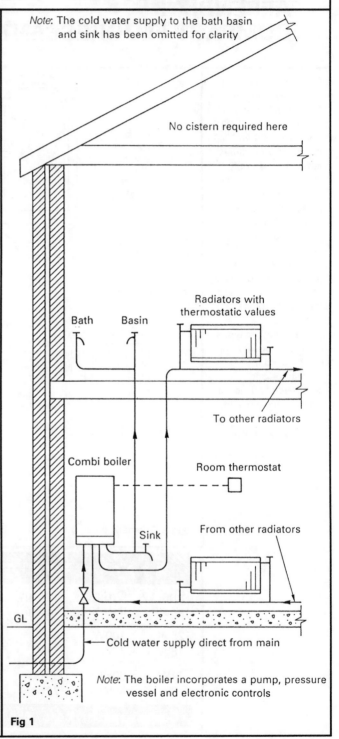

Note: The cold water supply to the bath basin and sink has been omitted for clarity

No cistern required here

Radiators with thermostatic values

Bath Basin

To other radiators

Combi boiler

Room thermostat

Sink

From other radiators

GL

Cold water supply direct from main

Note: The boiler incorporates a pump, pressure vessel and electronic controls

Fig 1

APPENDIX 5
GAS IGNITION DEVICES

Spark igniters are usually operated by mains electricity.

An electric charge is built up in a capacitor until a trigger device allows it to discharge suddenly and this current is stepped up by a transformer to a high voltage of 10 or 15 kV thus creating a spark. This spark is strong enough to light the main burner or the pilot.

Piezoelectric spark ignition consists of two crystals and pressure on them produces a large electric voltage creating a spark.

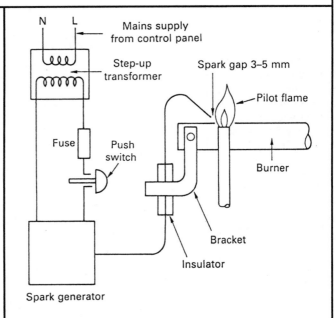

Fig 1 Mains spark igniter

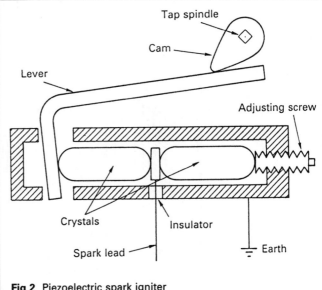

Fig 2 Piezoelectric spark igniter

APPENDIX 6   SMALL BORE PUMPED WASTE SYSTEM

A range of small bore pumping systems incorporating a macerator permits the installation of additional sanitary fittings almost anywhere in the building without the need of costly structural work or large bore pipes.

There are systems which will pump:

(a)   WC discharges horizontally up to 20 m

(b)   WC and wash basic discharges horizontally up to 50 m or vertically up to 4 m.

(c)   Discharges from an entire bathroom or shower room at distances given in (b).

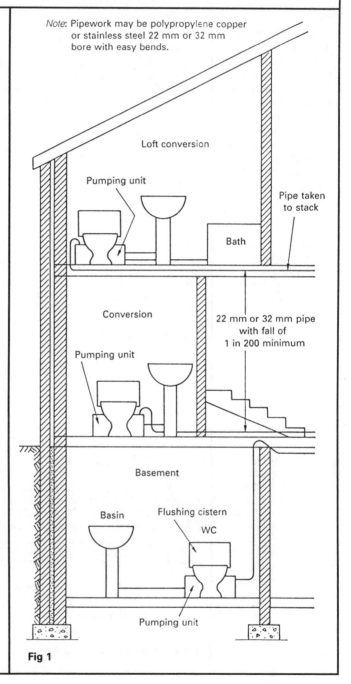

Note: Pipework may be polypropylene copper or stainless steel 22 mm or 32 mm bore with easy bends.

Loft conversion

Pumping unit

Bath

Pipe taken to stack

Conversion

Pumping unit

22 mm or 32 mm pipe with fall of 1 in 200 minimum

Basement

Basin

Flushing cistern

WC

Pumping unit

Fig 1

APPENDIX 7 LEGIONNAIRE'S DISEASE

Legionnaire's disease was first recognized in July 1976 when an outbreak of the disease occurred among delegates attending an American Legion Convention in Philadelphia, hence its name.

Various sections of the population are more susceptible to the infection than others, namely:

(a) Those over forty.
(b) Males.
(c) Smokers.
(d) Heavy drinkers.
(e) People having breathing impairment, e.g. asthmatic, bronchial.
(f) Those receiving treatment for cancer and related diseases.

Occurrence

The bacteria are common and can be found in many natural and man-made water systems. Soil is also a natural habitat.

A study in the UK found the bacteria (*legionella pneumophila*) in cooling towers, water systems, whirlpool spa-baths and clinical humidifiers.

Air washers and humidifiers with sprays providing very fine droplets of water can also be a source of infection.

These fine droplets of water, of about five microns in size, become suspended in the air and can be inhaled into the lungs.

Growth of the bacteria and prevention

The water temperature is the most important factor in the growth of bacteria. The micro-organisms can grow at temperatures between 20°C and 60°C but cannot grow at temperatures above 60°C. The bacteria are killed immediately when the water temperature reaches 70°C.

Legionella growth may be encouraged by the presence of:

(a) Stagnation.
(b) Sludge.
(c) Scale or rust.
(d) Algae and other organic matter.
(e) Muds and similar waste matter.

The removal of sediment from water systems reduces the risk and water systems must be properly covered to prevent the entry of dirt and vermin.

Cisterns must not be too large and if there is more than one cistern, water should run through each cistern to prevent stagnation of water.

Hot water dead legs should be reduced to the minimum as these may harbour the bacteria. Water left behind in shower equipment can harbour the bacteria. A drain valve can be incorporated which is designed to automatically drain residual water from the equipment.

Research has found that copper pipework tends to reduce the growth of the bacteria and boss white and hemp should not be used, but polytetrafluorethylene (PTFE) tape is recommended for jointing purposes.

Symptoms of the disease

Legionnaire's disease usually takes from two to ten days to incubate after exposure and for symptoms to develop. These symptoms are usually chills, headaches, high fever and muscle pains. Soon after, a dry cough and difficulty in breathing may occur.

In addition, about one third of the people suffering from the disease have diarrhoea and vomiting attacks, and about 50 per cent may become confused and delirious.

Legal requirements

The Health and Safety at Work Act 1974 covers the risks from the disease by persons employed in areas where there may be possible infection.

Outbreaks or suspected outbreaks of the disease must be reported to the Medical Officer for Environmental Health, Consultant in Communicable Disease Control, who advises the local authority.

If an outbreak of the disease occurs, which is the result of negligence, the persons responsible may be liable to prosecution by the authority.

Records for water systems

A maintenance record of water systems should be kept of the cleansing and disinfection of equipment. Managers should inform workers of the risk and supervision must be adequate.

Conclusion

It must be accepted that *legionellae* will probably be present in many water systems.

The task is to take all precautions to prevent its growth and possible outbreak of the disease especially in hospitals and old people's homes.

Although not everyone dies from catching the disease, in some outbreaks 10 per cent or more do.

Index

221